LIFE ON THE ROCKS

ALSO BY JULI BERWALD

Spineless:
The Science of Jellyfish and
the Art of Growing a Backbone

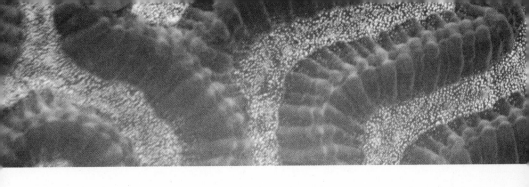

LIFE ON THE ROCKS

Building a Future for Coral Reefs

JULI BERWALD

RIVERHEAD BOOKS   NEW YORK   2022

Riverhead Books
An imprint of Penguin Random House LLC
penguinrandomhouse.com

Grateful acknowledgment is made for permission to reprint the following:
"Coral Restoration Is Now" lyrics copyright © 2018 by Scott F. Heron and Nathan Cook.
Used with permission.

Image credits: pp. 1 (and p. v, top), 31, 91, 157 (and p. v, middle),
221 © Alan Powderham/photos courtesy of Alan J Powderham;
pp. 113, 245 (and p. v, bottom) © Richard Vevers/The Ocean Agency;
p. 179 © FUNDEMAR/Sergio D. Guendulain-García.

LIBRARY OF CONGRESS CATALOGING-IN-PUBLICATION DATA
Names: Berwald, Juli, author.
Title: Life on the rocks : building a future for coral reefs / Juli Berwald.
Description: New York : Riverhead Books, 2022. |
Includes bibliographical references and index.
Identifiers: LCCN 2021026535 (print) | LCCN 2021026536 (ebook) |
ISBN 9780593087305 (hardcover) | ISBN 9780593087329 (ebook)
Subjects: LCSH: Coral reefs and islands. | Coral reef conservation.
Classification: LCC QE565 .B465 2022 (print) | LCC QE565 (ebook) |
DDC 333.95/53153—dc23
LC record available at https://lccn.loc.gov/2021026535
LC ebook record available at https://lccn.loc.gov/2021026536

Printed in the United States of America
1   3   5   7   9   10   8   6   4   2

Book design by Daniel Lagin

For Isy and Ben

Contents

PART VII.

WASHINGTON, D.C.

PART VIII.

AUSTRALIA, FROM AFAR

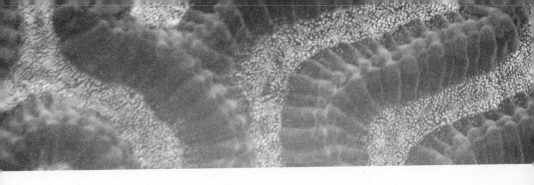

Part I

Reef Futures

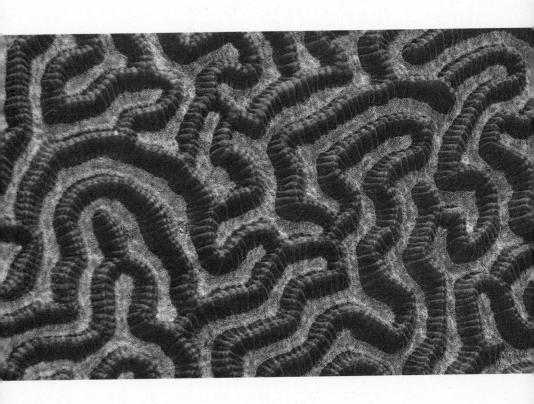

1

Fairy Land of Fact

It was love at first sight, for my part anyway. I'm pretty confident the corals felt nothing more than the waft of a current rolling off my flapping fins as I struggled to control my movements. But from the moment I dipped my eyes beneath the surface of the balmy Red Sea and kicked a few meters out to the reef, I was smitten. I had entered a world in which the sea gods and goddesses had conspired to mastermind a magnificent playground and then outfitted it in extraordinary decor. Awash in color and texture, the reef was beyond Baroque, more complex than Gothic. It was floral, it was animal, and it was mineral too. Each delicate petal and tendril was a revelation; each filigree and lattice an astonishment. It wasn't just my ineptness with a snorkel that literally choked me up. I felt emotional, overwhelmed by the simple recognition that this coral reef existed on the same planet as me.

What really made the reef so resplendent was that there was no sea divinity behind its magnificence. It was, as William Saville-Kent, the Great Barrier Reef's first Western biographer, wrote in 1893, a "fairy land of fact." The fairyland was the accumulated work over the eons of hundreds of thousands of tiny animals—most no bigger than the tip of a pencil—and the symbiotic algae that lived tattooed in their tissue. These creatures had none of the organs that we recognize as animal-like, no limbs or eyes or even brains with which to concoct this symphony of splendor. And yet, they had extraordinary capabilities. They were architects who designed the intricate structures of the reef. They were manufacturers who created the rock scaffolding of their homes. They were chemists who made their own protective sunscreen and complicated venoms. They were entrepreneurs who traded in the currency of nitrogen and carbon. They were soldiers who defended their territory from encroaching parties by firing poison-laden darts with unparalleled speed. They were hunters who used those very same extraordinary weapons to sustain themselves.

What was even more inconceivable was that these tiny beings were so much more than just their individual powers. And it was for the collective that my admiration of corals blossomed into true love. They were generous, sharing their nutrition with their neighbors through stomachs that were physically connected together. They were hospitable, building caves and dens for fish and crabs and octopuses and sponges. They were sensual. In the light of the moon, they spawned as one, releasing eggs and sperm upward in a deluge of synchronized hope for the future.

In the years following that first amorous dive on the reef, I changed my life in very significant ways, as one does for a true love. As often happens with passion, it didn't always go smoothly. But after many missteps, I did go to graduate school to study marine biology. Once there, I signed up for every chance I could to dive on other reefs: the Great Barrier Reef in Australia, and on the reefs surrounding Bora Bora, Jamaica, Maui, and the tip of Baja California. When I tucked my head underwater, the rush of love for the coral reef would always wash over me. Again and again, I was enthralled and entranced by the corals, by their creativity and synergy, by their beauty and complexity.

Until I wasn't.

More than a decade ago, I fell off the academic path and slipped into a career as a freelance science writer mostly working on textbooks, although I occasionally wrote for magazines and websites. My grandmother, who was in her midnineties, decided to throw a big party for herself because, as she wisely recognized, "you can't take it with you." She invited our extended family to join her on a Caribbean cruise. While I knew this voyage would be different from sailing on a research vessel, I was eager to see the vast horizon again and for the chance to dive beneath the turquoise waters in the Bahamas and swim around the coral gardens. This cruise company owned an entire island there and we were promised a day of snorkeling.

Aside from being at sea, the cruise was, as expected, strikingly different from life on a research vessel. On scientific cruises, work continued around the clock, which usually meant no more than a few hours of sleep at a time and a constant feeling of grogginess. Here, the ship's staff built a schedule to maximize our enjoyment of various shore activities. We sailed at night, rocked to sleep by the gentle roll of the waves, and awoke to a fresh new vista ripe for adventure each morning. The day we docked on the private island, I pulled back the blinds to the sight of a stunning double rainbow that ended at the beach. How they managed that feat, I had no idea.

The cruise company had thoughtfully supplied rugged wheelchairs with dune-buggy tires, so we could wheel my grandmother down to the beach and into the shallows. Once buoyant in the tepid water, she felt a freedom and lightness that age had stolen from her, and threw her hands in the air in happiness. As I held my toddler-aged daughter, Isy (short for Isabelle), in my arms, we bobbed around our matriarch in a kind of familial dance, basking in the sun and splash.

Afterward, while Isy dug holes in the sandy beach with her cousins, I collected my mask and snorkel to explore the reef. As the water deepened, I started to see small collections of silvery fish, dashing back and forth in the surge. But when I swam closer, I noticed that their scales were damaged and cloudy. Some even had blistery sores, open wounds on their flanks. As I reached the reef itself, if you could call it that, I saw only broken and displaced piles of rubble. Brown strands of slime streamed out from what used to be branches and boughs. As I swam on, I noticed absence. There were no urchins, no sea stars, no tubeworms, not even sponges. I didn't see shrimp crawling on surfaces. I didn't see crabs or snails crawling into crevasses. I looked under an overhang, where I

expected to see a few squirrelfish—crimson red, big-eyed, and antsy—
dart away. The cave was barren. Snotty algae grew everywhere like hunks
of moldy carpet. I lifted my head above the water, not wanting to see any
more. It was disgusting down there. Rather than a riot of color and texture of life and diversity, it was all slime and decay. I felt dirty.

I knew that the constant pressure of hundreds of cruise passengers
every week would take its toll on coral health and that the region had suffered hurricanes. I was also aware that coral reefs were in failing health
around the world. I'd read about bleaching and even written about spreading coral disease. But I hadn't experienced the rot and ruin until that day.
I didn't know what a dead reef *felt* like.

The truth is, there's no pot of gold at the end of the rainbow because
there is no end of the rainbow. The drops of water in the sky bend and
reflect sunlight to form a complete multicolored circle. You have to have
perspective to see the full rainbow. You have to get up high. A few mountain climbers and pilots have been able to take pictures of the circular
rainbow. Most of the time, we are too close to the ground to see it; most
of it remains invisible. But if we could have that perspective, we'd see that,
like so many things, the rainbow always returns back to where it started.
I should have never expected that the reef beneath that cruise ship would
end in a pot of gold. But it did bring me back to where I started.

A DEAD CORAL REEF ISN'T A RARITY. Today's coral reefs are assaulted by
a host of environmental stresses. The largest is climate change, which is
warming marine ecosystems even faster than those on land. Tropical
corals are the only corals that build extensive limestone reefs, but those
tropical corals live uncomfortably close to their upper temperature limit.
When they overheat, the algae that live within their tissues knock on
the thermostat and, finding no relief, submit their notice of resignation.
Or maybe the corals lay them off (much more on that to come). Sometimes corals and algae recover from this breakup—what has come to
be known as bleaching—but often they can't. Seventy-five percent of

the reef-building corals in the world have already been damaged by high seawater temperatures, which are on average 0.8°C (1.4°F) warmer now than in the twentieth century. Many have not recovered and are already dead. Half the Great Barrier Reef's corals have already died. For most, another degree warmer could be fatal.

The extra carbon dioxide we're adding to our atmosphere by burning fossil fuels doesn't just heat the air and water. When it mixes with ocean waters, it also lowers the water's pH, pushing the seas toward more acidic conditions. While the pH isn't yet low enough to dissolve coral skeletons, it is predicted to reach that threshold by 2085, if not before. How corals will respond is critical to their survival. Then there's overfishing, sedimentation from coastal erosion, ship anchors leaving scars, pollution from pesticide runoff and untreated sewage, unrelenting oil spills, and ever larger hurricanes. It's such a treacherous world for corals today that there are very serious predictions from very serious scientists that the world's great coral reefs may not exist by 2050.

What's at risk? A lot. While coral reefs take up less than 1 percent of the ocean's area, a fourth of all marine species depend on the reef at some point in their lives. A billion people rely on those ecosystems for sustenance or work. The combined revenue from food, recreation, and protection from storms attributed to coral reefs has been calculated at between $2.7 and $10 trillion a year. The death of the reefs means food insecurity for tens of millions. Corals are the most effective buffer known between the land and the sea, diffusing 97 percent of wave energy. A recent study by the U.S. Geological Survey valued the flood protection from coral barriers for U.S. citizens at more than $1.8 billion annually. Globally, the number increases to $9 billion. And that doesn't include their cultural significance or the yet-to-be discovered medicinal cures on the reef. That doesn't even include the joy of diving on coral reefs and celebrating their extraordinary beauty and diversity. But even more than all of that, the very idea that we might extinguish one of the major and most vibrant ecosystems on earth, that we might reduce hundreds of millions of years of evolution and tens of thousands of species to rubble, means we have a moral obligation to look with open eyes at what we've done and what we're doing. Given these dire predictions about coral's future, it seems to some that the only story left to write about the reefs is an obituary. But that's not just premature; it diminishes the story of coral to a quick headline. The ancient oath to love is "in sickness and in health," and so I resolved to look at the coral's sickness and see if I could find examples of healing. In September 2018, I put a Google alert on my email for the daily roundup of "coral reef" news.

AROUND THE SAME TIME, I found that it was impossible to contemplate the sickness on the reef without also considering the growing sickness in another of my loves, this one in my own home. During eighth grade, my daughter, Isy, changed in ways that were unexpected and frightening. She'd always been a kid with many friends, but about halfway through the school year, Isy abruptly distanced herself from them all. Snooping on her cell phone, I saw bubble after bubble of unreturned texts. She didn't respond to jokes. She ignored requests to hang out. She blew off invitations for sleepovers. When her friends resorted to the ancient form of communication, telephone calls, she didn't pick up. She brushed off my questions with a new defensiveness in her voice that I couldn't manage to circumvent. At the same time, Isy was spending more time in the bathroom. Within a few weeks, I noticed a line at her wrists, a demarcation between healthy skin and skin dried and tortured from repeated handwashing. Academically, problems were growing too. I saw missing grades in the online gradebook. I'd watch her do homework, but the teachers marked zeros—they said they'd never received her work. She'd study for tests but fail them, and not just a little. A 20 percent wasn't unusual. Her shame and frustration at this new person she was becoming was palpable. At one point, as I sat next to her at the dining room table while she struggled to write an essay on *To Kill a Mockingbird*, she'd yelled, "I just want it off me!" She collapsed on the floor, waving her arms around as if taking off an invisible coat. I was filled with horror and a sense of powerlessness.

NOT LONG AFTER I SET UP the "coral reefs" Google alerts, an announcement for a meeting in Florida called Reef Futures rolled into my feed. It was billed as not just about coral biology or about ideas for conserving the reef but for restoring it. Reef restoration was similar in theory to restoring an old home, but with coral colonies as the bricks and planks. When a reef was in disarray, people were learning to actively grow and then plant corals, refurbishing the undersea rather than just waiting for time and nature to fill in the void with a new life—as has been the practice espoused for decades by conservationists. This kind of rebuilding nature has ethical baggage associated with it, starting with the human hubris to think we can do it; passing through What if we muck it up worse? and skirting the corner of Can we really make a difference? and finally landing on Is it worth the cost?

I occasionally checked the Reef Futures 2018 program as new speakers were added. One day, I recognized the name of an Australian scientist named Daniel Harrison. I had a loose connection to Dan because he had worked in the same lab where I did my PhD. We were, as scientists sometimes say, academic siblings. Like me, Dan wasn't a coral biologist, and his talk wasn't directly about coral. It was about the coral's underlying problem: the warming climate. And he was proposing an audacious idea. Dan was slated to talk about building an ecosystem-scale air conditioner to lower the temperature of the ocean around the Great Barrier Reef. The idea belonged in a field known as geoengineering. Geoengineering ideas weren't new, but they have long been considered last-ditch efforts, ideas no one would seriously consider unless we had to. Were we at that point?

I hit the registration button.

2

Crazy Ideas

Driving off the tip of the peninsula that is Florida, you hit a series of small islands known as the Keys. The first and largest is aptly named Key Largo, and Reef Futures 2018 was held at its northern end in an exclusive community of yachters and golfers. Along the island's coastal highway, I watched the transitions from Key Largo's laid-back town of dive shops and pizza joints to the dense greenery of a chain of state parks and finally to a massive gate where uniformed guards scanned my driver's license before admitting me to the manicured grounds beyond. Winding my way through palm-lined streets, I parked next to immaculate tennis courts and found my way to registration and then to the chandeliered hall where the meeting was getting under way.

Two large screens flanked the stage and a giant banner proclaimed "Reef Futures 2018," uplit in purple and blue. The lights dimmed and a kaleidoscopic vision unfurled simultaneously on both screens: a tight zoom on the white-tipped tentacles of a pink coral waving elegantly in the surge. The video faded to the green corrugated ripple of a brain coral. Background music boomed and then settled. A quiet chorus of flutes peeked through as the video turned to high speed and churning tissue in parrot greens and cardinal reds danced across the screens. At the conclusion, the audience applauded boisterously and then was welcomed by one of the co-chairs of the Coral Restoration Consortium, which hosted the meeting. He told us that we, the assembled five hundred, hailed from forty different countries. Among our ranks were researchers, members of the media, philanthropists, and donors. I surveyed the crowd, seeing seemingly equal numbers of women and men, and many people of color. This coral reef community was rather more diverse than a typical academic meeting crowd. But philanthropists? Donors? That was unexpected. This meeting's intrigue was building. As I scanned the program, an entry caught my eye: at 11:50 a.m., a block of fifteen minutes was scheduled for a "Special Announcement!" What could that be?

Back on the stage a man named Tom Moore, the other co-chair of the Coral Restoration Consortium, was being introduced. Dressed in sleek black, he looked much more TED Talk than lab-coated scientist or blue-jeaned field biologist. He had an engaging and ebullient energy that radiated off the stage into the audience. Tom began by recounting many of the benefits of coral reefs, their outsize influence on marine ecosystems, their ability to protect those of us on land, their contribution to tourism, the spiritual and cultural contributions, and the untapped biochemical richness. But the bad news about bleaching was stark. To emphasize the point, he brought up the most severe example: half of the Great Barrier Reef—the largest biologically built structure on our planet—had been killed by warming water. But this *wasn't* the canary in the coal mine, Tom said, his voice strong in the swanky ballroom. A dead canary shows what isn't supposed to happen, a death no one thinks will happen or they wouldn't go into the mine. This death: it was predicted. The scientists in this room expected it. This mass death on the Great Barrier Reef "shouldn't be a surprise to anyone," Tom said.

I sat up in my chair. I had read about the massive bleaching of the Great Barrier Reef in 2015–2016 and then again in 2016–2017. I *had* been surprised. But it was becoming clear that the people sitting with me in the ballroom were already way past surprise. They were way past words like *preservation* and *conservation*. They were tossing out the environmental-ethical quibbling and asking, What can we do, and do now? To answer that, Tom explained, the coral restoration scientists had looked for other instances when animals had been brought back from the brink: pandas and whales. Asking the people involved in those struggles for advice, the coral scientists heard the same thing over and over: Don't be cautious.

Either the mood in the room shifted or my own mood shifted—maybe both. This wasn't a typical science meeting at all. Scientists are generally conservative by nature. The scientific method demands the accumulation of information before drawing conclusions. These scientists were saying, "Not in this case. There's no time for that."

Tom flipped through slides of coral farms: orchards of PVC pipes hung with coral fragments, metal stands studded with coral branches. He spoke of coral restoration scientists developing corals that could withstand repeated exposure to warming temperatures. He said scientists were developing breeding programs for corals that could survive the future. "We have had success at a local level," Tom said, "but there is a massive gap between small-scale coral gardening and success at the ecosystems level. And we don't have the technology to cross that gap yet." What was needed was a scaling up of local efforts, he said. But how to do that was the great unknown. It was an engineering problem that hadn't yet been solved. And that was what this meeting was about. "What we need is crazy ideas."

Tom didn't mince words summing up the stakes. "The status quo is losing the reefs. Climate change mitigation is still losing the reefs." What he meant was that, like an eighteen-wheeler coming to a stop, even if we do put the brakes on climate change, the accumulation of heat and carbon dioxide in the atmosphere will still ooze into the ocean for decades. It takes time to slow down that chemistry and physics. And by then it could be too late for the corals if we don't help them.

"Crazy ideas," echoed around in my head. Crazy ideas. This meeting wasn't an obituary for the reefs. It was about imagining a future in which reefs flourished. It was about inspiration. Maybe it was even about hope.

OVER THE NEXT HOURS, I listened to local stories of coral reef restoration. Scientists from Israel's Red Sea, from Belize, and from Curaçao told of their successes and struggles. Then Tom Moore took the stage again. According to the meeting schedule, it was time for the Special Announcement!

Tom opened, not with an underwater photo, but with a slide of an antique biplane, the *Spirit of St. Louis*, the name in iconic swooping brown font emblazoned down its side. A hundred years ago, on May 22, 1919, a New York hotelier offered a $25,000 prize to the first person who could fly nonstop from New York to Paris. A number of pilots died attempting to win the purse until, eight years after the announcement, an unknown mail pilot from the Midwest took off from Roosevelt Field in Long Island. After thirty-three hours, Charles Lindbergh landed at Le Bourget Field in Paris. The barometer he carried proved that the flight was uninterrupted. Spurred by the cash prize—a substantial one—Charles Lindbergh birthed the age of transatlantic flight and the massive aviation industry that followed.

The next slide wasn't an underwater photo either. It was a spacecraft. On October 4, 2004, the first nongovernmental organization successfully launched a reusable spacecraft into space twice within two weeks. Prior to 2004, it had been illegal for a nongovernment agency to launch a spacecraft into space. The rules had been changed for this spacecraft to fly. Like Lindbergh's flight, a substantial prize had been offered: $10 million by a group of technology entrepreneurs, including the founders of Google, who collectively called themselves the XPRIZE Foundation. The founder, Peter Diamandis, is often quoted as saying, "The world's biggest problems are the world's biggest business opportunities." What was perhaps even more important than the competition itself was the birth of a new industry. The XPRIZE drove the investment of more than $100 million in new space technologies.

"Big challenges," Tom said, "are driven by big prizes." With that, the lights faded and a video appeared of an even bigger and more elegant ballroom than the one we were sitting in. Two months earlier, tech entrepreneurs, CEOs, musicians, and politicians from around the world gathered in Los Angeles to consider the next XPRIZE challenge, a process called

"Visioneering." According to the website, Visioneering is XPRIZE's "vehicle for designing prizes that solve humanity's Grand Challenges. By tapping into the genius of the crowd, our global brain trust of philanthropists and innovators, we significantly increase the likelihood that our XPRIZEs will catalyze breakthroughs that generate a 10x impact in the world." On the stage, a jury sat in large gray armchairs. One by one, teams pitching the world's greatest big problems took the stage. They envisioned a future in which a billion more people had food, earthquakes were predictable, farmers were pulled from poverty, and the coral reefs could survive warming waters. After some winnowing, teams were pitted head-to-head: feeding a billion people versus building a global thermostat; extracting carbon from the atmosphere versus finding lost children. These were impossible decisions to make. "This isn't about picking winners and losers," the official from XPRIZE said, "this is about creating priorities." Audience members used an app to vote. Decisions were made by collective wisdom.

Two different groups speaking on behalf of coral reefs reached the finals. Both had big plans and narrow focuses: One group wanted to increase survival rates of newly planted corals. The other proposed farming the reef at a rate of half a billion transplants per year. Tom narrated: No one really expected either coral reef team to win. But then, in a last-minute pact-forming maneuver, the two coral reef teams joined forces, merging their pitches, offering up a bigger, more cohesive, and nearly impossible challenge: Saving Coral Reefs.

A man in the audience stood. He was a designer at Nike, who worked on shoes with stars like Kevin Garnett and Serena Williams. He said, "If we do not fund this prize, your grandchildren will *never* be able to find Nemo."

It was convincing. When the votes were counted, Saving Coral Reefs won in an unexpected landslide. The purse would be $8 to $10 million. Uproarious applause from the audience in the video in Los Angeles and the audience around me in Key Largo filled the room. Later, Matt Mulrennan, the director of the Ocean Initiative at XPRIZE, would tell me that key to the calculus was that the other projects, important as they were, had already had money thrown at them and still hadn't made much headway. Coral restoration had never been given the financial backing to give it a real shot. They saw huge potential for impact.

At the video's conclusion, Tom introduced Matt Mulrennan to the stage. Young and handsome in a tightly tailored suit jacket with a small red X pin on his lapel, Matt said the selection of Saving Coral Reefs at the Visioneering meeting was just the start of the XPRIZE process. The next step was to define the game. If you're giving out a prize, you need rules. And that's why he'd come to this meeting, to get insight from the experts. Matt posed the questions that needed to be asked: How do you define Saving Coral Reefs? How much area do you need to restore? Where does it need to be: one spot or many? How many species? How fast? Do you count only corals, or do you count the myriad fish and other invertebrates that live on a reef and need it to survive? How long does the coral need to live? How do you compare ecosystems across the planet?

As he spoke, my excitement began to sag under the weight of reality. Sure, you can hold a fancy meeting in L.A. with lots of wealthy, smart, inspired people. And, sure, you can have grand ideas to solve Grand Challenges. But then you confront our big, complicated, powerful planet. You confront the unpredictability of biology and the unrelenting march of development, climate change, ocean acidification, and pollution. You could put a price on it, but what did Saving Coral Reefs really mean? Was it even possible?

Up on the stage, Matt Mulrennan was all optimism and confidence. "We're planning to launch the XPRIZE in 2019," he said.

AS I'D FOLLOWED THE NEWS about coral reefs before the Reef Futures meeting, I had been saddened to read that one of its icons, Ruth Gates, had been lost about six weeks before. Ruth had been at the helm of a large and productive coral lab at the Hawai`i Institute of Marine Biology working on the molecular biology and genetics of corals. But perhaps even more important than the science she'd accomplished, Ruth was known for her relentless optimism in the face of the coral's bleak future. She was revered as a mentor to the dozens of students who passed through her lab. Ruth had been diagnosed with cancer five months before she passed away from a complication with surgery at fifty-six. The first day of the Reef Futures meeting, a video tribute to Ruth had been played, her infectious smile lighting up the screen. That evening, a spontaneous open mic turned into an appreciation of Ruth's life. One after the other, scientists took the microphone. In voices deep and high, in accents bearing the marks of the global reach of Ruth's influence, colleagues and students spoke of a mentor and an inspiration who would be deeply missed, a bright spot in a darkening field extinguished much too early. I wondered, as I listened to the speeches, if the undaunted positivity I'd begun to feel at this meeting was in some way an attempt to keep her spirit alive.

3

The Issue of Scale

The second morning of Reef Futures 2018 was handed over to a contingent from the Great Barrier Reef. Faced with sustaining the world's largest reef, the Australian government had mobilized a task force called the Reef Restoration and Adaptation Program, or RRAP. The program director began, "We Australians didn't want to be here. Restoration wasn't ever on our agenda. We wanted to just do conservation and preservation." In 1975, the Great Barrier Reef Marine Park Authority had been established to ensure that the reef was protected from destructive mining and fishing practices. Although it wasn't a perfect bulwark, the reef was widely hailed as one of the best-managed marine protected areas in the world. But not even the best local efforts could protect the Great Barrier Reef from the planet-scale heating, for which the Australian government itself holds some responsibility as the world's third-largest exporter of fossil fuels. "In the last two years, we've had to change fast," the RRAP director said. "With eighteen hundred miles of reef, it's going to take a lot more than scuba divers and baskets full of coral fragments [to make an impact]." The Australian government had provided an initial $6 million for a scoping project and then an additional $100 million for research and development to figure out what it would take to engineer the Great Barrier Reef's future. It was a "no stone unturned" approach. The scale of the project was astonishing: an ecosystem larger than Italy. But the stakes were enormous: the loss of not just a national treasure, but a global one.

Next up was the lead coral geneticist, Line Bay. Despite her Australian accent, her blond bob revealed her Swedish birthplace, where her first name was pronounced "Leena." Based on her work and that of other researchers, including Ruth Gates, Line's goal was to find corals with the genetic makeup to survive the coming high-temperature, high-acidity conditions. Even during mass bleaching events, a few corals always survive, Line said. Those animals were prime targets for her research. She also took advantage of the natural thermal variability in the ocean, selecting corals from the warmest seas that have already adapted to higher temperatures. Early experiments looked promising. Line and her colleagues had confirmed that corals could inherit the ability to withstand future hot conditions. "So," Line said, "we can theoretically breed hardier corals. But how do we propagate them and deliver them to the reef?"

Line flashed a slide on the screen that looked like the branching synapses of a primitive organism, color-coded in rainbow hues. "This is a mind map of what we think needs to happen," she said. When I looked it up later, I learned that mind maps are business tools, visual diagrams that contract and expand as more steps are added. They have been taking the place of PowerPoint to share complicated ideas. The sprawling diagram was the plan for what would need to happen to plant the Great Barrier Reef with future-hardy corals. It included developing husbandry programs, testing the animals for hardiness, growing vast numbers in nurseries, planting them en masse, monitoring growth, and freezing eggs and sperm in case of failure. It meant designing facilities in which to do all the work, developing logistics to transport the creatures to the reef, and many, many items in type too small to read from my seat in the audience. Line had made her point. This was an enormous undertaking. I sat back in my chair. The task seemed like nothing less than a moonshot.

Next to speak was the person I'd originally come here to see. Dan Harrison's work was part of the "no stone unturned" Australian project. Because the underlying cause of the dying reef was the changing climate, and his job was to study—audacious as it sounded—fixing the climate, or at least the climate over the reef, Dan showed a map color-coded with the average change in ocean temperature. Along the entire length of the Great Barrier Reef, the water had already warmed 1°C. The water was predicted to warm another degree, the brink of coral tolerance. "And then, what happens if we get an El Niño event that adds an extra degree or degree and a half? That would push the temperature into uncharted territory." Engineering the climate might sound incredible, but in the face of the uncharted, incredible was the only choice. "So, yes, I am serious," he said.

To engineer a reef chiller, one of two main options exists: cooling or shading. For cooling, possibilities include pumping cold deep water onto the reef or mixing warm shallow and cold deep water using giant fans. Both options are viable only if the cool water remains around the reef and doesn't get washed away. Computer simulations showed that pumping or mixing did result in some cooling, but the effect didn't last long enough to make a difference. Dan and his team concluded that neither pumping nor mixing was worth the cost. The second option was shading, much as we raise an umbrella on a hot day. One type of shade proposed was a powdered film spread over the sea surface, something like ground-up chalk that wouldn't harm the reef. Tests showed that a chalk film could block up to a third of the sun's rays. Another idea was deploying giant shade cloths, like canopies. Both of these ideas required a huge amount of shading material that would need to be replaced often—shading just wasn't scalable.

The most promising idea, Dan said, was adding reflectors to the clouds. Water droplets of the right size—about a micrometer in diameter—are really effective at reflecting the sun's rays. The sun's photons naturally bounce off a fraction of those tiny droplets and return back to space, where they can't warm up the earth. Cloud water droplets form around something solid like a scrap of dust, which is called a nucleator. Over the ocean, clouds contain about half as many droplets as clouds over land because there's just more dust over the land. But what if you could create more nucleators over the ocean? The answer is in the sea itself: salt crystals. Forcing salt water through thin nozzles, like the end of an airbrush, can create tiny droplets of the right size to reflect sunlight. If you make a big array of nozzles, kind of like a snowblower full of airbrushes, you create a nucleator-making machine. And the nucleator-making potential of seawater is astronomical. Dan's tests showed that sixteen billion nucleators could be made from just a teaspoon of seawater. The technique goes by the optimistic name "cloud brightening." Yes, changing the climate sounded dramatic, Dan agreed. But as everyone in the ballroom well knew, we'd already done something dramatic to the climate by spewing carbon dioxide into the atmosphere for the last century. The question was, would we do something dramatic to reverse it? "There are few interventions that are scalable to the whole reef," Dan said at the end of his talk. Cloud brightening was clearly on the table.

AS WE HEADED TO THE LUNCH BUFFET following the Australians' presentations, there was no denying the collective upbeat energy. Up to this point, coral restoration had been the Wild West, one restoration scientist told me as he passed salad tongs. Everyone had been figuring out how to grow and plant corals on their own, in their own ways. But here scientists and volunteers from around the world were meeting for the first time and sharing what they'd discovered. Ex-marines who dove together in Colombia shared tips with volunteers from Belize who had extensive experience with large-scale coral nurseries. Coral farmers from the Seychelles compared the reef stabilizers they used to the structures used in Sulawesi. The air teemed with positivity and potential.

And then I met some scientists from the Florida Keys. Holding my bamboo plate—no single-use plastic allowed at this meeting—I asked if I could join a group already seated together at a round table on a long, shaded porch. I was welcomed and shook the pawlike hand of John Hunt. With a mop of graying ginger hair and an easy smile swallowed up by full cheeks, John told me he'd been working in the Florida Keys for more than thirty years. Early on he specialized in spiny lobsters, versions of the more iconic Maine species minus the giant claw, that roam tropical coral reefs and depend on their nooks and crannies for habitat. But through his long career, he had also been involved in many conservation programs and in the establishment of marine preserves. When I asked about his recent work, John's eyes grew sorrowful. "Right now, it's the reef tract disease. It started four years ago up near Miami. And it's marching its way down the reef."

Despite my toe-dipping into coral science, I hadn't heard anything about this. John told me no one really knows what's causing the disease, and as he described its effects, the soundtrack of a horror movie began to play in my head. "It's a tissue-wasting disease," he said. "It eats away at their bodies. And it's nonspecific. It kills all kinds of corals, twenty-two different species. Only the branching ones are immune." I'd later hear the pathology of the disease likened to Ebola, the human disease that ravaged parts of Africa, causing massive cell death and organ failure. The lack of specificity was most worrying. Typically, diseases attack one species. Some, like rabies, are capable of jumping to several species. But twenty-two was about half the coral species in the Caribbean.

John suggested that I listen in on a session dedicated to the disease and its progress. For two hours, I watched scientist after scientist open their talks with the same slide, a map of the Florida Reef Tract, the third-largest barrier reef in the world, stretching three hundred miles from St. Lucie north of Miami down to the Dry Tortugas trailing off the tip of Florida into the Gulf of Mexico. The zone that the disease had already destroyed, a swath that abutted most of Florida's east coast and included about two thirds of the reef, was painted blood red. The next section, just south of Key Largo, was painted yellow, indicating the epidemic zone, a region where the disease was raging. There, it was, as one presenter said, "raining disease down on the corals." An invasion zone, where the disease had just been observed, stretched beyond that. Finally, at the very southern end of the reef, was a vulnerable zone where the corals were sitting in wait. The plague was headed their way. I saw photos of corals wracked with the disease, holes cored into their tissue that engulfed, in the time lapse of successive photos, entire colonies in just days. Tens of thousands had been killed. One scientist told of watching as nearly two hundred colonies in his study site—all more than three hundred years old—died one after the other in a matter of weeks. I saw photos of scientists trying to treat the corals using antibiotics embedded in clay and pressed onto the wounds like a poultice. Others tried epoxy mixed with chlorine. They returned to the wounded corals, day after day, week after week, tending to them like devoted nurses.

At the end of the heart-wrenching session, feeling despondent, I found John Hunt again. "What's going to happen?" I asked him.

A consortium of Florida agencies had joined forces, he said. As a collective, they were requesting funds to remove tens of thousands of still-healthy corals from the infected ocean. The consortium would house them in whatever facilities could be found until the disease front passed through. It was a massive undertaking. There were a lot of unknowns. But it was their best choice if they wanted to save the genetic and species diversity on the Florida reef.

I was incredulous. "Like an ark for the corals?" I asked.

John nodded in his sad way. "We are treating the corals in Florida like rhinos or cricket sparrows, like they are endangered animals." The consortium was a land-based Noah, saving creatures from doom in the sea.

I couldn't wrap my head around it, the unbelievable scale of the problem: the combination of the virulence of the disease, its unknown nature, as well as the attempted response.

"If you would have told me that I'd be treating corals lesion by lesion when I started working on the Florida reef, I would have laughed," John said. I'd seen the photos. Thirty years ago, the Florida reef was a place so bountiful it would have been impossible to imagine its state of decline today, impossible to imagine a time when a single sore on a single coral colony would matter so much.

I asked him how he kept going, wondering how it would feel to see the corals you have known and dived among for decades faced with an epidemic.

"I think we've all gone through a period of grieving," he said. "And if I don't do everything I can, then fifteen years from now, if I'm still alive, I'll regret it."

4

Buying Excitement

Just after lunch on the last day of the Reef Futures meeting, I sat in on a session called "Success Stories." The meeting organizers had culled projects from around the world that showed some of the best progress. I slipped into the back of the hall just as a team from Puerto Rico was wrapping up their story. Next up was a young man with an English accent who introduced himself as "Jos, from Mars."

"Not that Mars," he added after the joke landed. "From the Mars, Incorporated, the company." Yes, he was talking about the makers of some of your favorite Halloween treats. Jos proceeded to let photos speak for themselves. He showed an image of a dead coral reef, shattered and gray. It was from Sulawesi in Indonesia, he said. There blast fishing, or fishing using explosive bombs, has had a devastating impact on the reef.

Jos clicked forward on his presentation. "And this is two years later." The screen lit up with a vibrant reef full of color and life. Jos pointed to a small part of the reef where an interlocking bit of rebar poked through, looking like a six-rayed star. Jos said that Mars had developed a simple, transportable technology based on rebar structures to which fragments of broken corals were attached. The structures were networked on coral rubble to form the infrastructure for a new reef. Within a year coral cover increased by as much as 70 percent. "We're not just planting corals," Jos said. "We're building reefs."

This did look like success. I highlighted the project in my notes and put a star next to it. I wanted to learn more from Jos from Mars.

IF THERE IS ANYONE who might be considered a celebrity in the coral reef world, it's Richard Vevers, star of the documentary *Chasing Coral*, which has been showing on Netflix since 2017 after an award-winning premiere at the Sundance Film Festival. It's probably not an exaggeration to say that *Chasing Coral* changed the way more people understood the plight of the reefs than the collected research papers published by the scientists who sat in the room with me. We like visuals, and we like stories. Richard Vevers gave us both.

Richard was slated as the final keynote speaker of Reef Futures 2018. His talk began with a photo of a roll of toilet paper, which drew the expected chuckle from the crowd. This, he said, in a charming British accent, was his job on the day he walked out of a London advertising agency, ending his decade-long career. During a discussion about ways to better sell four-ply over three-ply, he'd realized that there just had to be more to life. He decided to couple his expertise in advertising with his longtime love of diving, establishing a nonprofit called The Ocean Agency to bring attention to the declines he'd seen on the reefs over the years.

A major issue, Richard recognized, was that the corals are hidden; often invisible. You have to get wet to see them. How could he get them noticed? Richard approached Google Earth and asked if they wanted to develop the equivalent of Google Street View for the underwater world. Google agreed the idea was a good one but asked Richard to find a partner, which he did in an insurance company called XL Catlin. Because XL Catlin's business was risk, they asked him to tie the project to their mission. Richard then asked for input from coral reef scientists, who explained that traditional reef surveys were slow and painstaking, requiring counting and mapping individual corals in a square-meter plot of reef. Richard and XL Catlin realized the first step to assessing risk is knowing what's there in the first place. They could establish a baseline by developing more sophisticated cameras and using machine learning to stitch scenes together and analyze species composition. Together, the perhaps unlikely partnership of Google, an insurance company, coral scientists, and Richard's The Ocean Agency compiled the first large-scale photographic survey of the Great Barrier Reef in 2012. On the day the underwater street view was posted to Google, "more people went virtual diving than have ever been diving in history," Richard said.

Richard had shrewdly emblazoned Catlin's name on the black housing of every camera used in the project. Despite their protestations that black was the wrong color for working in the tropical heat, he convinced the survey crew to wear a uniform of sleek black Catlin T-shirts. Richard was building a brand around coral reefs. The media picked up on it. The underwater street view project garnered a spread in *Time* magazine as well as thousands of other media pieces. Richard estimated the advertising benefit to Catlin at $200 million, which well exceeded their investment. Google had a new internet tool to tout. The coral scientists were on their way to having coherent baseline surveys of the world's reefs. The public could view coral from the comfort of their living room couch. Richard seemed to have found a winning, optimistic story.

And then the ocean had its say. It was 2017. About a third of the way through additional surveys of the Great Barrier Reef, the water warmed, a lot. Corals started bleaching. Richard and his team pivoted. With their powerful cameras already on location, the crisis was a critical opportunity to do what he'd set off to do when he walked out of the meeting about the ply of toilet paper: increase the public's awareness of the plight of the reef. Richard dubbed the situation not just another bleaching event, but the Third Global Bleaching Event. Attaching a number to the situation made all the difference in terms of capturing attention, he explained. It made people wonder how they'd missed the first and the second. And yet, they felt that they were still on the cutting edge if they were party to the third. The clever branding caught the attention of documentary director Jeff

Orlowski, who had made the successful film *Chasing Ice*, about the shrinking ice sheets around the world. Orlowski flew to the heart of the bleaching in New Caledonia.

There's an unforgettable scene in *Chasing Coral*. The team is organizing their gear in a makeshift film studio that they've rented in a side room of a party boat. Outside on the deck, a DJ plays music while drunken tourists dance above the overheating reef. The driving music sounds something like "hunts, hunts, hunts, hunts." Richard Vevers and the crew thread their way among the oblivious revelers, lugging tanks, fins, and cameras, and splash nearly unnoticed into the water. Beneath, the corals are mustering poorly understood biochemical pathways to cook up a kind of natural sunscreen in an attempt to fend off the excess radiation that threatens to roast them alive. The colors of this self-made balm look unnatural, artificial. They are highlighter shades: fluorescent yellow, neon pink, and day-glo indigo. They are painfully bright and visually stunning. Except for the occasional skeletal white of an already dead colony, the fluorescence blankets the entire sweep of the reef, reflecting against the mirrored sea surface, blurring and swirling into a kind of morbid tie-dye.

The dying coral reef, Richard explained, falls in the category of a "wicked problem." The theory of wicked problems was first proposed by social policy researchers Horst Rittel and Melvin Webber in the 1960s, while working on what seemed to be intractable issues in government and society. One characteristic of a wicked problem is that it has several explanations. This is true for dying coral: warming water, habitat destruction, pollution, illegal fishing, and more are reasons why corals are dying. Also, wicked problems are actually symptomatic of another problem. This is also true for corals; dying corals are symptomatic of the underlying problem of climate change. And wicked problems are hard, perhaps impossible, to claim success at solving. This is true for corals as well: Can't a reef always be healthier? Psychologically, what wicked problems do, Richard said, is engage the logical part of our brains in a very different way from the creative part. The logical part can understand the problem and respond to its urgency. But the creative part really wants to feel good. How do you convert a hard, logical problem into a positive feeling? Well, said the ad man, that's what branding is all about.

It's already happening, he pointed out. Big brands like Adidas are signaling their concern about ocean health, making shoes out of recycled ocean plastic. Method soaps use recycled ocean plastic for their bottles. Customers drive this "brand standing." Richard said, "Nine out of ten millennials will switch brands for a cause." He had teamed up with a design firm in London, the computer software giant Adobe, and the Fortune 500 financial management company Accenture to explore what it would mean to develop a brand for the coral reef. They'd all been captivated by those day-glo colors in *Chasing Coral*. There was magic, unreality, and beauty in the colors of the reef's struggle. So the team's first shot across the bow was a palette—highlighter yellow, neon pink, day-glo indigo—that represents the fragility of the reef. Richard flashed mock-ups of familiar logos—the striped Adidas, the crowned Corona beer, the encircled Body Shop—on the screen, all branded in the coral palette of brilliant inflorescence.

Was it possible? I thought. *Could these colors do what Richard hoped? Would they be capable of making coral reefs visible in a way that was otherwise impossible? Would this palette bring corals out from the ocean's depths and onto our clothes, our shoes, our beer labels, maybe even across our faces in our eyeshadow and lipstick? Would we wear coral clothes, walk in coral shoes, drink from coral cans? Would they start a movement?* Maybe *this* was the crazy idea.

"This is the first conference I've been to where there is actual excitement about the future of the coral reef," Richard concluded, squaring his shoulders to the audience. "People buy excitement."

RICHARD VEVERS'S KEYNOTE was billed as the last talk on the Reef Futures 2018 agenda. Afterward, Tom Moore, who had kicked off the meeting, returned to the stage to reiterate how this week had been about cooperation and hope. He thanked all the sponsors and presenters, gave instructions for getting to the reception at the Caribbean Room, a beer joint he described as "the opposite of the Reef Club" with its chandeliered ballroom and uniformed staff. "But before we get to the beer," Tom said, a couple researchers had asked for just five more minutes of our time. "They promised me it would be worth it."

Two scientists claimed the stage, one of them wielding a ukulele. The slide on the screen switched over to a photograph of a holiday tree topped with a star. For the last few days, I had been so consumed in the world of coral that I'd all but forgotten that it was mid-December and Christmas was around the corner. But as I looked, I recognized that the tree wasn't an evergreen at all, it was made of branching coral and PVC pipes, like the orchards for farming corals beneath the sea. The ukulele struck up a familiar tune about Santa Claus coming to town, and the two scientists on the stage were gesturing for the crowd to stand. Along with five hundred others, I felt myself rising as I sang along to the words that appeared on the screen.

We've made up a list,
It's tough we must strive
All working together
For reefs to survive
Coral Restoration is Now!

It's larval propagation
Transplanting corals too,
Assisting evolution
All these goals we must pursue!
Yeah!

In front of me, people's heads danced back and forth; smiles bloomed on faces as they pulled out their cell phones to record this bizarre, joyful moment. The slide switched and the song went into the second verse.

You better watch out
The urgency's high
Let's raise up a shout
And make it a cry
Coral Restoration is Now!

For the people so reliant
We must instill belief,
Let's engineer the future
A living, vibrant reef.
Yeah!

A laugh rolled across the room during the pause: a collective release at the absurdity of singing a modified Christmas carol about coral. But the plucking wasn't stopping; we were headed back into the final chorus.

Oh. You. Better watch out
The urgency's high
Let's raise up a shout
And make it a cry

Coral Restoration . . .
Driving Innovation . . .
Coral Restoration . . . IS NOW!!!

Yes, it was silly. Yes, it was kumbaya. But it was also the simple fun of sharing music and melody with people who cared deeply about the planet and its health. I thought about the coral stories I'd heard. Examples of loss and ruin from bleaching and the horrible disease ravaging Florida were tempered by the undeniable excitement of restoration efforts like those from the Mars project in Indonesia and the massive initiative in Australia. The promise of the XPRIZE might even be enough to shift the needle, not only in the public's perception, like the kind of solution Richard Vevers was describing, but also in on-the-reef know-how. The applause and cheers that erupted at the end of the song buoyed the feeling of something big and important and maybe even consequential all around us. More remained to the story of coral reefs than an obituary.

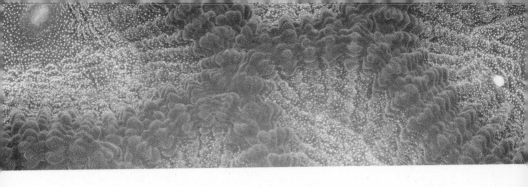

PART II

WHAT KEEPS A
REEF TOGETHER

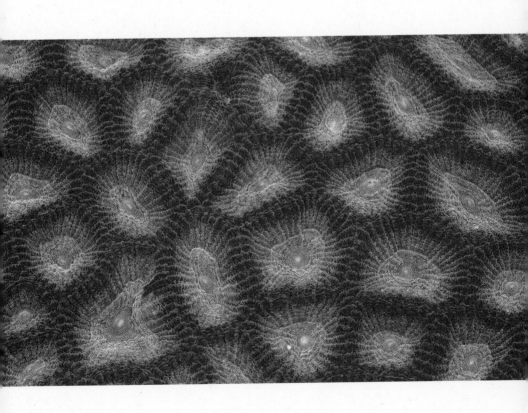

5

A Badass Merger

few weeks after Reef Futures 2018, I clambered up the steps of the biology building at The University of Texas at Austin and thought about how so many biology-building stairwells have the same echo, an acoustic created by metal flooring and walls of layered, chipped paint. Pulling open the heavy fireproof door on the fourth floor, I paused to read one of the many posters that are the artwork in these kinds of halls, creations left over from meetings past. The poster explored the genetics of a disease from the Flower Garden Banks National Marine Sanctuary, where coral reefs grow atop subsea salt domes in the Gulf of Mexico. Out of the corner of my eye, a man with close-cropped hair wearing an orange shirt and white linen pants came into my view from a side hall.

"You must be Misha," I said.

Misha Matz, the principal investigator of the coral genetics lab at The University of Texas at Austin, nodded in the affirmative. I had arranged to meet Misha to get an academic perspective on the plight of corals. As someone who studies coral genetics and evolution, he could provide the long view of how populations might be able to adapt to our changing seas. Plus, I could drive down the road rather than fly across the world to meet him.

"I'm just reading the posters." I gestured.

"No one ever does that," he said in a warm but also slightly dismissive tone that I would come to recognize over the following years. "Come," he continued, nodding down the hall he'd emerged from. "I'm Russian. So we must have tea."

As I followed Misha toward his office, I noticed a circular tattoo right in the middle of the back of his neck, a sort of filigreed mandala. Later, I saw the same logo on his lab website and realized it was the six-sided skeleton of a coral polyp. Even later, I would discover that Misha had once been in a rock band in Moscow.

One corner of Misha's office was furnished with chairs and a coffee table, on which rested a bowl of chocolates. A teapot hidden under a quilted tea cozy the shape of a chicken rested nearby. Its old-fashioned hominess felt a little out of sync with Misha's rock-and-roll vibe, but it was also inviting.

After tea was poured and chocolate unwrapped, Misha said, "So, you want to hear about my unpopular ideas?"

"What makes your ideas so unpopular?" I asked.

"I think most proposals out there are just bizarre. And they have no scientific justification."

As he spoke, I started to believe that it wasn't exactly that Misha's ideas were unpopular, but that Misha had more than a healthy dose of skepticism and believed everyone else should as well.

"Most people start their talks by saying, 'Evolution is too slow, it can't possibly catch up with the speed of climate change.'" Misha gestured to an article on a side table next to me, a preprint of a recent study. I picked it up and looked at the map on the page. It showed 680 reefs dotting the Indo-Pacific Ocean. Lines connected the reefs like spiderwebs, showing how coral larvae traveled on ocean currents. The probability of each coral reef's survival was predicted based on genetics and future water temperatures. In the model, the ocean was programmed to heat due to global warming. The corals were programmed to evolve to the shifting temperatures.

Misha took a bite of chocolate, perhaps to heighten the dramatic pause.

"I cannot make them go extinct. I'm sorry, it's impossible. There doesn't seem to be any theoretical limitation to their evolutionary potential."

While many scientists were concerned that corals couldn't evolve fast enough to keep pace with our planet's heat, Misha was saying that at least for the corals in his model, evolution had already equipped them with the tools to survive their warmer future. While I wondered what that meant for restoration, Misha was already racing on to the next set of questions: If a coral had the genes for surviving higher temperatures, would it be compromised in another way, like having a weaker immune system or an inability to survive in a more acidic future sea?

"We looked into this," Misha said, "and we find no trade-offs whatsoever. Instead we find reinforcements. Whoever is badass is badass. Which means you just select for badassness on any axis of variation, and you get a nice, future-adapted, robust coral. There IS such a thing as general badassness."

I laughed because, really, what could sound better than coral badassness?

EVEN BEFORE MISHA ARTICULATED the idea that corals were badass, I already had my suspicions they were. The most critical piece of coral's badassness is the powerful, and still quite mysterious, symbiosis between corals and the algae that live within their tissues. This badass merger has built structures more voluminous than the Pyramids of Giza or Teotihuacán and longer than the Great Wall of China. When Alan Shepard took off from Cape Canaveral, Florida, on May 5, 1961, in the *Freedom 7* Mercury capsule, he deployed his periscope four minutes into his flight and beamed these words back to mission control: "What a beautiful view . . . I can see Okeechobee . . . identify Andros Island . . . identify reefs." The rocky skeletons that corals build are famously the only biological structures on our planet visible from space. And what's perhaps the most badass is that it happens where it shouldn't be possible at all. Coral reefs are constrained to tropical realms, places with lots of sun to power the symbiotic algae. But as opposed to the richness of tropical rain forests, these tropical seas are nutritional deserts. In a bucket-sized sample of tropical seawater, you'll maybe capture a few dozen plankton. In a bucket-sized parcel of seawater from the middle latitudes, you'll find ten times more life. The reason for the difference has everything to do with the heat that makes the tropics feel so different from the temperate regions. Warm water floats on the colder water below. The oceanographic term for the transition from the warm surface to the cold deep sea is the thermocline, and in the tropics it rests about three hundred feet below the surface. Similar to the transition you can see with vinegar floating on oil in a bottle of salad dressing, it is a physical barrier. When living things die or release wastes in the surface layer, all that organic matter sinks down through the thermocline into the deep, like the oil part of the salad dressing. In that dark, deep sea, like in the soil of your garden, bacteria compost those wastes and carcasses, creating fertilizer.

Consider what it would be like to be a single-celled alga floating in the sea. This alga is capable of performing photosynthesis, which is a way of saying it's in the solar power business. The alga's solar panels are chlorophyll molecules. Its storage batteries are sugar molecules. The alga needs supplies to build and maintain both its solar panels and its batteries. Those supplies aren't silicon wafers and lithium ions, but rather carbon, nitrogen, and phosphorus: the elements in fertilizer. The existential problem for a tropical alga is that to perform photosynthesis, it needs both light from the sea's surface and the supplies for solar panel parts, which are buried beneath the thermocline. How is a minute alga supposed to connect the two when they are separated by at least a football field's distance? On land, plants solve this problem with a long stem that connects sun-drenched leaves to nutrient-bathed roots. That's not an option for a single cell. One strategy is to live in places where strong currents mix the deep and the shallow water, but those places are rare in the tropics. Another strategy is to swim, migrating vertically from the deep to the shallows every day. But that takes a lot of energy. So one innovative group of algae turned to the ecological marketplace. And they found a partner looking to cut a deal.

Now, shift your perspective and imagine a baby coral newly planted on the seafloor. It has begun construction of its skeleton, and the fleshy animal part—the polyp—extends its baby tentacles into the seawater for the first time, maybe waving them about a bit like the fingers of a jazz dancer. There's a bit of unknown magic involved in what happens next. Maybe a small bit of food, perhaps a tiny crustacean, bumbles by while making a buzzing sound. Stinging cells in the coral's tentacles, the same type of micro-injectors found in jellyfish, respond to the noise and smell of food and deploy. Once they hit their prey, the stinging cells inject toxins for good measure. The tentacles curl around the injured crustacean, folding it into the coral's mouth, and from there into its stomach. After the coral digests its first dinner, it's left with a bit of nitrogen, which is just what the baby coral needs to do some advertising for the solar power systems it needs to build.

The coral posts a chemical HELP WANTED sign by exuding nitrogen, the kind of chemical currency that attracts algal applicants, which it then ingests into its stomach. Or perhaps the coral got lucky with its first meal. Perhaps the poor plankton has recently eaten and in its stomach are a few algae. Or it may be that the coral baby is washed over by bits of sand, and attached to those grains are a few algae, which it inadvertently swallows. It's even possible that a fish swims by and poops on the tiny coral. In its poop, there's a bit of algae. However it happens, rather than making a meal of the algae in its stomach, the young coral forms a merger with them.

The first thing the coral does is build a kind of work cubicle for the alga inside one of its own cells. The walls are a bubble of cell membrane technically called a symbiosome. Into this bubble, the coral releases a chemical perfume that signals that the alga no longer needs to bother with the cumbersome problems of commuting around the sea or avoiding predators. In response, the alga casts off its swimming flagella, as if hanging its jacket on the coatrack. The alga then creates a supply list for the coral. First, the alga needs sunlight, but not too much and definitely not the damaging kind from UV rays. The coral obliges by living near the sea surface, orienting its tentacles to capture light, and producing mucus and other proteins that act as UV sunscreens.

Next, the alga needs carbon dioxide—a lot of it. The coral can supply some of that carbon dioxide through respiration. Like us, the coral is an animal, and it uses oxygen to break down its food, in the process producing carbon dioxide. But that won't be enough. In order to fill the alga's hefty order, the coral constructs special pumps engineered to pull carbon dioxide from the surrounding seawater. Also, the alga needs nitrogen to build and maintain the cellular machinery—like chlorophyll—used in photosynthesis. With these materials the alga can finally make a product that the coral wants desperately: sugar.

This is how the coral and the alga become ideal trading partners: the alga powers the coral by making sugar, and the coral provides the supplies the alga needs to make that power. In one recent study, just six hours after the coral was fed plankton, it transferred almost a quarter of the carbon and nitrogen from that meal to the alga. In exchange, alga provides more than 90 percent of the sugar from photosynthesis to the coral.

I'd be remiss not to point out that trades similar to those between coral and algae are the basis for most food webs on our planet. All plants use light and carbon dioxide to perform photosynthesis and make sugar. Animals everywhere burn sugar for energy, releasing carbon dioxide. This plant-to-animal-and-back trading goes on every day, every second, and everywhere. It is the crux of the cycle of life. What is so profound about the alliance between algae and coral is its intimacy.

Often the best parts of a science story are in the details, and that's the case here too. The exact nature of the relationship between coral and algae is under intense debate. Some see it as a mutualism: the coral and algae benefit roughly equally from their alliance. But others see it as a hostile takeover. The coral micromanages the nitrogen supply, providing just enough to keep the algae pumping out sugar, but not enough for the algae to become oversatisfied. If that happened, the algae might shift their focus from sugar production to reproduction. Some coral scientists call this scenario "milking the algae," evoking the way a dairy farmer milks cows. If so, coral might be thought of as the original cultivator and algae as the planet's greatest domesticated crop. But evidence is starting to show that algae have more say in the matter than cows do. Misha Matz, for his part, suspects that the algae are parasites, that they "milk" the coral for an easy supply of nitrogen and carbon dioxide. (More on that soon.)

I like to think about the badass merger as a partnership that produced not just the greatest construction project on our planet, but the greatest *green* construction project our planet has ever known. Here's my evidence: If the coral/algae merger were going to make a case for its success before, say, a board of directors, it might be asked to show how it stacks up against the competition. In tropical oceans, ecosystems typically produce similar amounts of organic matter as in deserts (about 0.2 grams of carbon per square meter per day). But on the coral reefs, the production of organic matter is a hundredfold greater, at 20 grams of carbon per square meter every day. That is equivalent to the maximum productivity of a very efficient agricultural crop, like sugarcane growing under optimal conditions. A wheat field in the rich soils of Illinois that is fed artificial fertilizers is ten times less productive than a coral reef. Roberto Iglesias-Prieto, who studies marine science at Penn State, made the case this way: "Together, corals and zooxanthellae [another word for the algae that live in coral] are the world's most efficient light harvesters—far better than plants. They can absorb the same amount of light as a green plant but with an investment of an order of magnitude less chlorophyll, which is the most expensive thing for a primary producer to create." The value of this unique and highly coupled merger would impress any shareholder.

That corals have been able to succeed where the seas should never have supported so much life led scientists to call coral reefs "Darwin's paradox." And, indeed, they are. Owing to the way we've interpreted Darwin, when we think about what leads to success in the biological world, the narrative is usually one of competition and survival. Similarly, in the world of business, we tend to talk with the same vocabulary: competition and marketplaces, of disruption and survivors. By contrast, the corals and their badass merger with algae is a story of collaboration and of outstanding results. It's a story that, in these divisive times, we'd do well to emulate.

6

Hopeful Monsters

That first conversation with Misha brought up a lot of ideas I wanted to explore further. Misha said that in his models the Indo-Pacific corals couldn't go extinct. Why not? My path to the answer started with a five-volume treatise with a not-so-humble title, *Description de l'Univers* (Description of the Universe), written by a Frenchman named Alain Manesson Mallet and published between 1683 and 1719. In its final form, it was a massive picture book of the continents, the heavens, and charming, if a bit cartoonish, drawings of cities like Mecca, Havana, and Tokyo. The folio included people from around the world too: the king of Persia adorned in golden robes and a cerulean cape with a twisted Salvador Dalí mustache; Greenlanders modeling fur jackets and boots; and Indonesians wearing flowing toga-like sarongs, flitting among greenery as if vacationing with the Greek gods. Mallet drew the idyllic images himself. A stint as a schoolteacher taught him the importance of keeping his audience entertained.

One section of Mallet's treatise, called "The Natural World," includes a drawing that I love so much I hung it just to the right of my desk. Titled "Von den Corallen," or "Of the Corals," the image shows an ocean scene busy with boats, flags waving off masts, and sailors traipsing around ropes and gear. As your eye scans the image, it unexpectedly settles and is held by a detailed pair of bare buttocks poking out of the water (Mallet's schoolkid tactics totally work on me). From there, your gaze drifts to the lower half of the drawing, where more bared legs kick frantically at the air, and you realize the sea is teeming with naked men. And then you notice that all of the characters who are not ass-upward hold in their hands branches of jagged coral. Aboard the vessels, corals are stacked like piles of firewood among the gear.

I originally found the image of the naked divers in the book *Corals of the World*, written by the modern grandfather of coral taxonomy, John E. N. Veron, and several coauthors. Veron's connection to nature was so great that while in school he was nicknamed Charlie, after that other Charles: Darwin. The name stuck, and the legacy it foretold was, at least in the coral world, accurate. Charlie Veron was the first full-time researcher on the Great Barrier Reef and the first scientist the Australian Institute of Marine Science hired. He discovered the great biodiversity of the Coral Triangle and is responsible for naming about 20 percent of all known reef corals.

Veron included the naked coral divers picture in his book to show that coral collecting was a big business for much of the seventeenth and eighteenth centuries. Unlike catching more motile creatures like fish, it was easy to snap off coral branches. Because of the hard skeletons, preservatives like those needed to keep other soft creatures from rotting were unnecessary. Back in Europe, coral skeletons were prized in natural history collections. But this chaotic collection system had a long-lasting impact on the science of coral identification. Few scientific details, like where and when the coral had been found, made their way back to Europe. Through the centuries, naturalists and museums loaned the skeletons to one another. What scant information was associated with those skeletons was often misplaced or lost, leading to two centuries' worth of misnaming and misidentification of coral species. Veron wrote, "Of the estimated 2,400 nominal extant zooxanthellate coral species in existence, 15 percent have no taxonomic record and those that do have one, have their names embedded in the vagaries of nomenclatorial history."

Besides what we humans have done to confuse ourselves, corals are themselves inherently tricky. Despite their being familiar to humans since antiquity—Perseus was said to have spilled the blood of Medusa, ossifying seaweed into the coral reefs of the Red Sea—it wasn't until 1726 that French physician and naturalist Jean-André Peyssonnel proved that corals were animals and not plants, as had been previously thought. It's not hard to see why. A polyp's rim of tentacles form a petal-like fringe around its disc-like mouth. And like flowers, the colors span the rainbow, and the patterns delight the imagination. Coral's first cousins, sea anemones, share that same floral anatomy and were named for the colorful rounded anemone blossoms that grace gardens across Europe. The scientific name for the class that includes corals and sea anemones is Anthozoa, or "flower animal." Adding to their plantlike nature, many anthozoans are solar-powered, getting most of their energy from the badass merger that feeds them the sugary fruits of photosynthesis. But even in the centuries since corals were firmly planted in the animal kingdom, they remain taxonomically squirrelly. They are shapeshifters and masters of disguise. Depending on light, food availability, and local currents, the same species can take the form of leaves, branches, or plates that look nothing like one another. Even within a single colony, the shapes of polyps and their underlying skeletons can vary dramatically. The two best ways to really identify a species are to cut a thin section of the skeleton and look at it under a microscope or sequence its DNA. Neither is possible when you see a coral out on the reef.

But all this messiness also makes me feel better. Even after studying corals for years, I often can't tell what's what. I'll look at a delicate whitish-blue coral in an aquarium and wonder if it's an *Acropora*, an *Anacropora*, or even a *Porites* that's gotten all pointy at the tips, or yet some other branching coral I've never heard of. And it's a bit comforting to know that even the experts have trouble. A genus I learned thirty years ago as *Montastrea annularis* is now known as *Orbicella annularis*, and it turns out that it's not just one species, but three. A species known as *Pocillopora damicornis* can look like anything from a head of cauliflower to a spindly haunted forest tree, depending on where it's living. And there are at least four other species of *Pocillopora* that are easily confused with *Pocillopora damicornis*, depending on whether you are on the eastern coast of Australia or in Hawai`i.

One big reason for this caprice is that there is a lot of hybridization among corals, with the gametes of similar species capable of fertilizing each other, and those larvae growing into offspring that then reproduce themselves. A common textbook definition of a species is "individuals that produce fertile offspring with one another." So if hybrids are being produced all the time, and those hybrids themselves produce fertile offspring all the time, does that mean each new hybrid is itself a species?

Yes, according to *Corals of the World*. Most of us are used to thinking of evolution taking the shape of a tree in which species diverge and diverge from each other, like boughs and branches diverging off a trunk. But for corals, the shape of evolution is more like that of the patterning of a python, which looks like a network of interlacing lines that pull away from one another and merge back together. Veron and his colleagues call it reticulated evolution.

Corals, being moored to the seafloor, must rely on currents to move around their spawn. The eggs and sperm of most corals are released into the water only once or twice a year, coordinated by the light of the moon, the setting of the sun, ocean temperatures, and other cues that I'll fill you in on a bit later. Once liberated, they are at the fate of the seas. If there are strong currents, these gametes may travel long distances. Populations might meet populations from far away, mixing up the gene pool and keeping the genetics of the species diverse and vibrant. But if the currents slow, as can happen when the seas rise and fall with ice ages and temperature shifts over geologic time, then that mixing can slacken. Then eggs and sperm mix only with locals. The gene pool shrinks. Genetic drift sets in. Coral species diverge. Now, suppose the sea's geography shifts once again. The currents strengthen, mixing the seas and the spawn it carries more powerfully. Gametes meet gametes they haven't been in contact with for millennia. Hybridization occurs. Species that had been divergent come back together, or as Veron says, they are "repackaged." This is the process of reticulated evolution: separation and repackaging, diver-

gence and convergence. "Species" is just a convenient term to characterize stable groups in a particular moment. That species change all the time
is the DNA of evolution. This idea of species being characterized by mor-
phing rather than stability echoes what Darwin had in mind when he
proposed "descent with modification." Corals, then, rather than an aber-
ration, are the very ideal of evolution.

Reticulated evolution is not a quirk of coral alone. Recent studies show
that there was plenty of hybridization among even our close ancestors.
Two so-called separate species of baboons in Kenya that diverged from
each other 1.4 million years ago have been mating with each other for
thousands of years, and their offspring are able to survive and reproduce
just fine. Even closer, Neanderthals are now confirmed to have interbred
with humans as recently as forty thousand years ago. By one estimate, as
many as 10 percent of all animal species interbreed to form hybrids. Scientists wonder whether these "hopeful monsters," as the hybrids are
sometimes called, might be evolutionary shortcuts, a quicker way than
traditional natural selection to add new, useful traits to a population.

Another important benefit of reticulated evolution, it is thought, is
that the organisms that exist in these weblike systems are not as prone to
extinction, that they are more resistant to major evolutionary shifts. For
the last seven thousand years of Earth's history, the climate has been
stable and the sea level has been largely constant. That, of course, is shifting. The sea's temperature has already risen by a degree. In the tropics it
is predicted to increase by half a degree Celsius per decade for the rest of
the century. By 2100, our seas are predicted to rise by three feet, but ten
feet isn't out of the question—and it's even possible these estimates are
too conservative. As the water rises, will enough light filter down to support the coral's algal partners? How will the changing currents impact the
movement of the coral's spawn? Will the badass merger remain intact in
some places but not others? The corals we see today are the success stories
of Earth's history, the fragments of a system of reticulated evolution that
have survived the massive upheavals in our planet's geological past. But
whether these hopeful monsters of the reef will be resistant enough to

withstand what's coming—and the speed of its arrival—remains in question.

ONE EVENING, my daughter Isy and I sat at the dining room table, her chemistry homework in front of us. It was a review of the parts of an atom. She had once known the information, but it was as if the knowledge had evaporated from her brain. Her pencil hovered above a page already streaked with the scars of past erasures.

I said, "So, you're supposed to draw a hydrogen atom. What are the parts?"

"Proton, neutron, and electron," she muttered in a flat, automatic voice, the kind kindergartners use to repeat the Pledge of Allegiance before they understand the meaning of the words.

"And which ones go in the nucleus?"

She looked down, her hand hovering above the paper. Her face was a kind of blank, a flat, rocklike expression. But beneath the surface I could feel the swirling, growing panic. Her brain was not processing the information. There was a battle raging inside. The storm transferred its energy to my own body: in my back, in my cheeks, in my shoulders. I placed my face flat in both my hands and dug my fingertips into my forehead, as if I could iron the furrows flat. I drew my head back on my neck and was surprised to hear the crack of my occipital bone against muscles pulled taut at the base of my brain where reptile impulses are formed, urges of flight and of fight.

And I wanted to do both.

7

Tiny Architects

In 1835, in Valparaíso, Chile, a young Charles Darwin was awakened from his sleep by the heaving of the ground beneath him. In the aftermath of a massive earthquake, the shoreline was lifted higher. Bands of mussels and barnacles were now exposed, like the rings of a drained bathtub. The rock layers beneath looked like the edges of a book, revealing fundamental truths about our planet. Darwin recalled a recent publication by the great naturalist Charles Lyell. It said that the continents weren't stuck in one place, that they could shift positions on the globe, that they collided and separated, that they lifted and fell. Layered on that idea, Darwin contemplated the coral reefs.

Dotted throughout the Pacific Ocean were rings of corals, called atolls, although Darwin hadn't yet sailed west of South America to see one. The prevailing theory, which Lyell proposed, was that atolls formed when corals encrusted the tops of subsea volcanoes, forming a circular rim. But looking at the raised shore of South America after the earthquake, Darwin imagined the whole continent lifting upward. He reasoned that in response to something rising on a fairly incompressible and spherical Earth, some*thing* else some*where* else had to be sinking. Push your fingers into an air-filled beach ball; nearby the ball pooches outward slightly. Darwin figured that if South America was pooching outward, the thing that was sinking somewhere else must be in the Pacific Ocean.

He wrote, "I may remark that the general horizontal uplifting which I have proved has & is now raising upwards the greater part of S. America & as it would appear likewise of N. America, would of necessity be compensated by an equal subsidence in some other part of the world.— Does not the great extent of the Northern & Southern Pacifick include this corresponding Area?"

Darwin knew coral grew in the shallows around islands. If the Earth's crust under the Pacific was sinking, that would mean that the islands themselves were sinking. And that meant that the corals had to grow upward to remain near the surface. Darwin suspected that the animals must climb toward the light by building their homes upon the backs of their ancestors' skeletons below. When the *Beagle* reached Cocos Island off Costa Rica a year later, and Darwin finally saw coral atolls, he wrote, "We must look at a Lagoon Isd [island] as a monument raised by myriads of tiny architects, to mark the spot where a former land lies buried in the depths of the ocean." As the land submerged, what remained of every island was a ring of sun-seeking coral where land had once been. Darwin's thought experiment proved largely correct, but he would never know it. A century later, in the 1950s, scientists developed the technology to drill through an atoll in the Marshall Islands as preparation for testing atomic bombs. Rather than a thin veneer of coral around a volcanic rim, they discovered the ancient remains of corals more than four thousand feet thick. As the seas had been sinking, the corals had been building their rock homes, molecule on top of molecule, for millions of years.

The reason Darwin's thought experiment didn't prove entirely correct was because he didn't yet know that for the last several million years, the seawater level has been rising and falling as much as a hundred meters, repeatedly. When the world is cold, seawater freezes in glaciers and the sea level falls, exposing coral to the air. Rainwater, which is more acidic than seawater, etches away at the exposed skeletons in a process called karstification, particularly in the centers of the reefs. When the earth warms, the glaciers melt and the sea level rises, flooding the coral once again. New corals grow atop the surfaces, bolstering the already slightly higher outer rim. Over time, the edges are repeatedly reinforced and the centers are repeatedly eroded to create the rings we see today. The thousands of atolls dotting our seas are a living memorial to the weathering from the skies above, the flooding of the seas below, and the persistent, upward growth of life.

ONE AFTERNOON, I yelled for Isy to come downstairs so I could drive her to dance class. "Are you ready?" I grabbed my car keys from the counter.

"Yes," she answered, walking toward the door dressed in shorts, a T-shirt, and tennis shoes.

Although her dance school didn't have a strict dress code, they did expect students to wear typical dance clothes like leggings and tank tops. "You can't wear that," I said. "Go put on your dance clothes. We're going to be late!"

"I'm fine."

"No, you can't dance in that."

"I can."

I let loose an ugly mother stare.

"I can't!" she screamed.

"What do you mean, you can't?" I yelled back.

"Then I won't go!" She ran to her room, throwing herself on the purple and blue flowered sheets she'd had since she was a toddler.

"We've paid for those classes." I stomped after her. "You are going! I don't care what you wear!"

She howled as she walked down the hall and threw herself in the car. I got in the driver's side and buckled my seat belt angrily. "I am not letting you lose dance," I said. "You've lost so much." And I fiercely pulled out of the driveway.

A few blocks away, through her tears, Isy said that her dance clothes were "contaminated." Wearing them would contaminate her too. This idea of contamination was an angry sore that seemed to be spreading through her life, and our family's. Certain people and certain places could contract "contamination." There was no logic to it. Sometimes I would drive down a street and Isy would go dark, begging me to turn around because we were on a route that would take us past a place that was contaminated. At the grocery store, she'd ask to stay in the car. We might run into a contaminated person there. Now, it seemed, the contamination was in our home too, folded in her dresser drawers. None of it made any sense, but she was unshakable in her conviction.

Later, when we told her psychiatrist about the episode, she said that the way around it was to expose her to the trigger, let her experience the anxiety it produced, and eventually she could learn to tolerate it. Anxiety can last only so long, and then it will diminish. It's like wading into cold water. You stop feeling it after a while.

That weekend, we planned an exposure to Isy's dance clothes. I handed her a "contaminated" tank top and leggings. She pulled them on, and it was like watching my child dip herself into hot oil. The clothes seemed to ooze pain onto her skin. She collapsed on the floor, sobbing and writhing. Any soothing my husband, Keith, and I tried was useless. An excruciating hour and a half passed before the pain and tears began to ease.

I didn't know enough about Isy's condition then to understand that exposures need to be performed in a hierarchical manner, starting with triggers low on the anxiety scale and slowly building upward. That way, you gather evidence that you can handle the exposures; you gain tolerance. We had forced Isy to jump into the deep end of the pool before learning to wade in the shallows. I know now that the technical term for what we put her through is "flooding."

THE BASIC BUILDING BLOCKS OF CORAL REEFS' rocks are molecules made of calcium and carbonate that the coral gather in dissolved form from the seawater. For the last million years or so, the seawater's pH has remained at about 8.2. At that pH, the ocean contains enough carbonate that animals who need it to form skeletons can easily find it in the water around them. But as burning fossil fuels adds carbon dioxide to the atmosphere, some of it seeps into the ocean—about twenty-two million tons each day—and lowers the pH. Today, the ocean's pH has sunk to around 8.1. If we continue burning fossil fuels at the current rate, the pH is expected to plummet to 7.8 by the end of the century, or even lower. At a lower pH, marine creatures that build calcium carbonate structures must use extra energy to find the carbonate building blocks they need. That ocean acid-ification compromises their ability to do other important things, like gather food and reproduce.

About a decade ago, just as the atmospheric CO_2 pushed against 400 parts per million for the first time, a threshold that hadn't been reached at any time in human history, marine scientists began raising alarms about ocean acidification. Early predictions put coral at risk when atmospheric carbon dioxide reached 450 parts per million, which is expected around 2050. Coral would undergo complete demise at 560 parts per million, which will occur before the end of the century if nothing changes. How-ever, when scientists tested coral at lower pHs, the impacts of ocean acidification on corals were ambiguous. Sometimes they found impeded growth. In other cases corals were unaffected.

Inside coral tissue, an incredible system of pH control—what I like to think of as the alchemy of light and stone—is at work. In the fluid right next to where a coral builds its skeleton, special molecular pumps simul-taneously move calcium ions into the fluid and pull hydrogen ions out. (The H in pH represents hydrogen ions, which control the acidity of any fluid: the more hydrogen ions, the more acidic, and confusingly, the lower the pH.) The result is that right next to the growing skeleton, there's a stew of calcium and carbonate at a pH that favors the formation of stone. In the process, the coral has removed a bunch of extra hydrogen ions that threaten to acidify the coral's own tissue. But the coral can use that hy-drogen in another location. By pumping the extra hydrogen ions into the algae's workspace bubble, it decreases the pH there, which conveniently favors the formation of carbon dioxide. The rich CO_2 supply is a big ben-efit for the alga because it needs carbon dioxide for its solar-powered photosynthesis. There's one last piece to the system: the coral's hydrogen pumps require a lot of energy. Where does that energy come from? From the sugar the alga makes from solar energy. Today, researchers believe declines in calcification may be only half as great as previously predicted because coral and algae already have sophisticated systems in place to compensate for an increasingly acidified sea. Misha Matz summed it up this way: "On acidification, don't let people tell you that it's a problem. Corals don't seem to give a shit. And that's good news."

While living coral can wage a chemical battle against the corrosive effects of an acid sea, a dead reef cannot. Clams, oysters, worms, and even seaweed dig into the calcium carbonate with thousands of tiny biological chisels. Those creatures excavate nooks and crannies where they can live while exposing more surfaces to erosion and corrosion. In September 2019, researchers released pulses of carbon dioxide to a study site on the Great Barrier Reef for two hundred days to mimic future ocean acidification conditions. They found that places without living coral eroded twice as fast. And because so much coral has died, we are already nearing a situation where millions of years of accumulated reef is dissolving faster than the remaining coral can build it. In Florida, the northern reefs are already eroding faster than they are growing, and the middle and southern regions are getting close. Some reefs seem to still be building in the Pacific, but predictions are that there, too, reefs will dissolve faster than they can grow by 2080. In his book *Reef Madness*, which chronicles Darwin's fight for acceptance of his theory of coral atoll formation within the scientific community, David Dobbs wrote that the coral created "with their calcified skeletons the huge structures that joined the organic and inorganic worlds as well as the sea and land. In the early nineteenth century, some saw in coral reefs a welcome antidote to the erosion that . . . was erasing humankind's terrestrial platform." Today, we are at risk of losing the formula for that antidote, but probably not because of ocean acidification. For corals, the alchemy of light and stone depends on the algae's solar power. When the temperatures rise and the algae abandon the coral, the incantation won't function. By 2100, it is predicted, under carbon dioxide emissions that would raise global temperatures by 4.5°C, and subsequent bleaching from the heat, no reefs will be able to grow fast enough to match the rise in sea level.

AS ISY'S CONDITION WORSENED, her therapist helped us understand that underneath her anxiety was a severe case of obsessive-compulsive disorder. OCD is not, as popular culture sometimes portrays it, a tendency to be organized and clean. It is not lining up your shoes and pencils in tidy rows. Isy certainly didn't care about any of that. Rather it is characterized by an obsessive thought pattern that gets stuck in your head playing over and over and ramping up anxiety. The release for that anxiety is to perform a ritual, or compulsion. The problem is that when you perform the compulsion, you only reinforce the idea that the compulsion works to alleviate the obsession. That creates an endless and ever-stronger loop of obsessions and compulsions. We all probably have some form of obsessive thoughts and small rituals we perform. I knock on wood to stop the anxiety that comes with saying a bad thought out loud or to avoid jinxing my luck. But when these thought patterns prevent you from living the life you want to live, it's a disorder.

OCD has many incarnations. For Isy, one was contamination, which originated with a compulsion for handwashing and long showers. It then spread to an obsession with certain places, people, and clothes. The contamination wasn't about germs or pathogens, but something ineffable and consistent only with the logic of her OCD. The compulsion that arose was to avoid them, which was why she limited her social life and movements outside our house so severely.

Another was scrupulosity, an overconcern with an ethical issue. Isy told us that the reason she had been scoring so low in school was because she was obsessed with lying and plagiarism. Her compulsion was to destroy her work in case she hadn't done everything entirely by herself. That's why she had stopped turning in her homework. It was why she intentionally failed tests. Eventually, she stopped reading, her fears about lying about what she'd read making it too difficult to acquire new information. For me, as a writer and a scientist whose life revolves around knowledge, Isy's OCD felt especially cruel, corroding one of the greatest values—the joy of learning—that I'd most hoped to pass on to my daughter.

8

Bleaching Beginnings

The story of how we came to understand today's dissolution of the badass merger of coral and algae, known as bleaching, begins at the westernmost of the chain of islands that dribble off the end of Florida, called Loggerhead Key. There, among the scrubby grasses, you will find a stone memorial with a metal plaque. It was placed there in 1923 and reads:

ALFRED GOLDSBORO MAYOR WHO STUDIED THE
BIOLOGY OF MANY SEAS AND HERE FOUNDED A
LABORATORY FOR RESEARCH FOR THE CARNEGIE
INSTITUTION DIRECTING IT FOR XVIII YEARS WITH
CONSPICUOUS SUCCESS BRILLIANT VERSATILE
COURAGEOUS UTTERLY FORGETFUL OF SELF HE WAS
THE BELOVED LEADER OF ALL THOSE WHO
WORKED WITH HIM AND WHO ERECT THIS TO HIS
MEMORY BORN MDCCCLXVIII DIED MCMXXII

Mayor traveled the world studying the animals of the seas, mostly jellyfish, but other invertebrates too. He conceived of Loggerhead Key as the site of one of the United States' great marine biology institutions, Tortugas Laboratory, and petitioned the Carnegie Institute to provide funding. Mayor was the station's champion and first director and, as the plaque explains, so utterly forgetful of self that he sacrificed his health in pursuit of science. One evening, after a day's work at the lab, he collapsed into the seawater. He had been ill with tuberculosis, which was untreatable in those days before antibiotics. Mayor's wife erected the memorial a year after his death.

Working at the Tortugas Laboratory and in Australia in the early 1910s, Mayor performed the first rigorous experiments on marine animals and water temperature. When it came to corals, Mayor found that most species tolerated an increase of 10°C (16°F) before they died. That finding was notable because jellyfish, which are close cousins to the corals, could tolerate much larger increases of as much as 16°C (30°F). Mayor made a prescient comment. "This low factor of safety in tropical marine animals may at times become of biological significance." Those times are now. Today's oceans have absorbed more than 93 percent of the Earth's warming over last 150 years. That heat energy is equivalent to the heat generated by more than three nuclear explosions every second for the past twenty-five years. A hundred years ago, Alfred Mayor understood just how little wiggle room there is for tropical corals when it comes to warming seas. But he didn't understand exactly why.

At the same time, another young scientist at Tortugas Laboratory was looking at how corals withstand other kinds of stresses, like changes in salinity and darkness. Thomas Wayland Vaughan would himself eventually direct another great American marine science lab, Scripps Institution of Oceanography in California. Vaughan's expertise in marine science ranged from seafloor sediments to coral to unicellular protozoans called foraminifera, but his academic interests were even broader. He also studied comparative religion, Teutonic legends, and Asian art. Once, losing patience while trying to identify a coral, he told a student, "If you continue studying corals, you will either become famous or lose your mind!" But Vaughan worked on corals anyway. His experiments showed that rather than immediately dying when exposed to fresher water, the corals first turned pale. Together with Mayor, Vaughan placed coral in the dark for as long as a month and a half. Again, the coral didn't die but instead turned pale. These are the first records of coral bleaching in the scientific literature. It took another decade before the connection between elevated temperatures and bleaching was made.

IN THE 1920S, the Royal Geographical Society of Australasia organized an expedition to study the biology of "the greatest of all coral reef formations." At that time, newly minted marine biologist Maurice Yonge was finishing up some postdoctoral work on oysters in England and looking around for his next gig. He was in a bit of a rush because he and his wife, Mattie, had recently married and money was growing tight. Mattie was something of a hotshot herself, having just received her medical degree. Together, the young duo made a compelling package of marine biological expertise and medical know-how to the Royal Geographical Society, which was sending a research team to a remote atoll where self-reliance would be critical. Maurice was hired to lead the expedition, which sent the "handsome couple" to Australia for a year to study the reefs.

After a long boat trip and some stops to glad-hand Australian officials, Maurice, Mattie, and about twenty other researchers set up a small study site on Low Isles, about two thirds of the way up the Great Barrier Reef. Their home and labs were nine small bungalows, which included sleeping quarters, a wet lab, and a kitchen/cafeteria. In a picture of the research party, Mattie and Maurice sit front and center. She has a short dark bob, parted to the side, and wears a white blouse and Bermuda shorts, her legs crossed at the ankles in a way that doesn't look demure at all. She leans slightly to the left toward her husband, who also wears a loose white button-up shirt and shorts, his dark hair brushed straight back above a high forehead. The newlyweds both squint in the tropical sun.

Though the researchers cobbled together a diving bell operated by a car tire pump to explore the reef from underwater, the contraption wasn't particularly easy (or safe) to use. That meant much of the work was scheduled around especially low tides when the water peeled itself back, revealing the coral. During one of those low tides in February 1929, Maurice noted that the water was unusually warm. He wrote that he was "impressed with the temperature of the water in the pools, which was literally hot to the touch."

One month later, during the low tide on March 21, Maurice again walked across the same reef flat. He wrote, "Great numbers of whitened skeletons of corals killed by the heat a month previously, were seen." Some of the white colonies, however, caught his eye. They seemed different and "on closer examination were found to be alive and apparently perfectly healthy, but with colourless, transparent tissues. They resembled in every way corals which had been living in the absence of light, and whose tissues consequently were almost or entirely without zooxanthellae." While Mayor had previously bleached corals by putting them in dark boxes and Vaughan had done it by changing salinity, these were the first corals observed that were bleached by high temperature. Maurice chipped off small samples and dropped them into jars full of a preservative so he could look at them more closely in the lab. Then he hammered stainless-steel bars into the reef to mark the locations of the bleached coral.

The next strong low tide was on April 11, and Maurice hiked out to the exposed reef again. He went back to those same whitened colonies where he'd left the stainless-steel markers. "All five were found to be perfectly healthy, the chipping off the samples having done them no apparent damage. They were all distinctly brown in color, although much paler than average colonies of the same species. Samples were again taken and fixed."

Maurice was unable to return to the reef during the low tides in May because he was on an expedition, but he did go back again during the low tide on June 3. "All five colonies had by this time resumed their normal deep brown colour, i.e., the zooxanthellae had apparently multiplied until they had regained their normal abundance within the tissues." Maurice collected samples once again. Back in the lab, the coral tissue was sliced into very thin pieces to observe under the microscope. In the first sample, taken when the corals were white, algae were present, but in very sparse quantities. He observed algal cells in the process of being ejected from the coral's tissue. Otherwise, however, the coral's cells looked normal. In the sample taken fourteen weeks after the heat wave, Maurice observed that the algae were as numerous and healthy as in a coral that had never bleached, perhaps even more numerous. He wrote, "There is thus evidence that, *under natural conditions* [*sic*], corals may not only be killed by high temperatures, but . . . may themselves survive although their contained zooxanthellae have been almost completely ejected. The question arises, are the algae directly affected, or are the corals so injured that the zooxanthellae are no longer able to live within them and so are ejected?" Ninety years later, incredibly, we still don't know exactly which partner—coral or algae—initiates the dissolution of the badass merger.

On the return trip to England, Maurice and Mattie stopped in New Zealand, Fiji, and Samoa, as well as Honolulu. They went on to San Diego, where they stopped at Scripps Institution of Oceanography and met with the director, Thomas Wayland Vaughan, who had done the original studies on coral bleaching and salinity with Alfred Mayor at the Tortugas Laboratory. No one knows if Vaughan and Maurice compared notes about their observations of coral bleaching, but the two pioneering coral scientists seem to have hit it off. Vaughan offered Maurice a job on the spot. Maurice declined, and he and Mattie returned to England, where he went on to a successful career as a marine scientist, eventually earning a knighthood. In 1942, Mattie suffered from a tragic brain illness, which took her life just three years later. As his academic career was beginning to wind down, Maurice wrote a comprehensive book of marine natural history called *The Sea Shore*. It is dedicated to Mattie, "who will walk no more on shores with me."

In 2019, ninety-one years after that first exploratory trip, two sets of couples traveled to Low Isles. Ove Hoegh-Guldberg and Sophie Dove and Maoz Fine and Efrat Meroz-Fine, who are among the most experienced coral scientists in Australia and in Israel, respectively, located the same spots that the Royal Geographical expedition studied and repeated the same assessments. Massive shifts had occurred. The temperature increased from an average of 26.6°C to 27.7°C. The acidity of the water decreased by a tenth of a unit, from a pH of 8.24 to 8.14. Sea level increased by more than twenty centimeters (8 inches). The reef had been pummeled by cyclones, bulldozed by infestations of the coral-eating Crown-of-thorns starfish, and ravaged by repeated bleaching events. In 1928, skeleton-building corals dominated the reefs; today, those same places are full of soft corals, which don't build a rocky reef and which provide fewer habitats for marine creatures. Among the hard corals that remain, the scientists found a shift from structurally complex branching corals to more thermally tolerant boulder corals. The size of those remaining hard corals shrank by as much as 30 percent. Despite much searching, many species of urchins, snails, and sea stars that Maurice Yonge and his team recorded were nowhere to be found. It's hard not to wonder whether Maurice and Mattie would recognize Low Isles if they returned today.

9

Bleaching Bombardment

After the experiments of the early twentieth century, the phenomenon of coral bleaching faded into scientific backwaters. The study of ecological processes was on the rise, and the reef was rich with relationships and complexities to unravel. With the widespread growth of scuba in the 1960s and 1970s, marine scientists were finally able to immerse themselves in the waters they studied rather than peer from the surface during low tide or dip in with nets. During those decades, they counted, identified, and measured. They used the term *zonation* to describe just how carefully different coral species calibrated their light and depth regimes, and how those zones created microcosms and microzones for countless other species. They studied how sand and sediment, algae and fresh water changed communities. They investigated interactions among fish, worms, sea urchins, snails, shrimp, crabs, and coral. What they found was a world more complex than they'd ever imagined. So coral biologists didn't think much about bleaching. Why would they? There was too much vibrant life to understand.

What brought coral science back to the question of bleaching was electricity. In the late 1950s, a surge in the demand for electricity called for an expansion of the power supply in Hawai`i. The Hawaiian Electric Company built a new oil-powered facility on the western side of Oahu, about eighteen miles from Honolulu. Like many power plants, it was situated on the shoreline so that it could use seawater to cool its turbines.

When it passed in the 1970s, the Clean Water Act required that the electric company develop a monitoring program to document how the discharged water, which could be as much as 6°C warmer than the surrounding seawater, affected the coral reefs.

Two young researchers at the Hawai`i Institute of Marine Biology took the monitoring job. Paul Jokiel had just moved to Hawai`i from Illinois on a teaching fellowship and was hoping to study coral. Steve Coles had just finished his master's in marine science at the University of Georgia and had come to Hawai`i for his PhD. They received a small budget and built a marine station from the ground up, even pouring concrete themselves. They figured out how to control the temperature in their aquaria using the heat released from the back of an air-conditioning unit.

Out on the reef, Steve and Paul defined levels of coral damage from the power plant's heated seawater as "dead, bleached, pale, and normal," what I believe was the first time the word *bleached* was used in the scientific literature. The corals nearest the outfall, where the water was 4° to 5°C warmer, were largely dead. Moving seaward, where the heated water mixed with the cooler local water, the corals shifted to bleached; and when the temperature was less than 2°C warmer than the surrounding water, the corals were largely unaffected.

When Paul and Steve had the opportunity to travel to Enewetak, in the Marshall Islands between Hawai`i and the Philippines, they found that corals could withstand temperatures around 2°C higher than the corals they'd been studying. Then again, the average water temperature in Enewetak was at least 2°C warmer than in Hawai`i. Over time, they extended their temperature studies to looking at reefs in Florida and in the Arabian Gulf. They worked out that the temperature tolerance of corals wasn't an absolute number, say 29°C or 31°C. Instead, it was relative. Corals could withstand a temperature increase of 2°C above what they were used to for a few weeks, wherever they lived. Otherwise, they would bleach. Remember that forty years earlier, Alfred Goldsborough Mayor discovered that coral could survive a temperature increase of 10°C in the Caribbean. But his exposures were just a few hours, whereas the

coral near the power plant outfall experienced the increased temperature continuously. Paul and Steve's surveys revealed that over the long
term, corals' temperature limit was about 2°C above average before they
bleached. Today, the amount of temperature increase and time exposed
has been formalized into the term "Degree Heating Weeks," which is
calculated by multiplying the increase in temperature by the number of
weeks that a coral experiences the increased temperature. At 4 Degree
Heating Weeks, you can expect bleaching. At 8, you'll see widespread
bleaching and mortality. But back in the 1970s, the question of how much
heat a coral could withstand and how long before it bleached or died
seemed kind of a niche question. The temperature tolerance of a coral
mattered only if the coral had the bad luck of living in front of a power
plant's hot water outfall or happened to be on the top of a reef during a
very low tide during a particularly hot summer.

That all changed in 1980.

In the Florida Keys, the currents usually sweep up from the south,
bringing warm water but also creating gyres, huge spinning eddies that
suck deep, cool water upward. In the summer of 1980, however, the normal currents lay stagnant. Clear, cloudless skies prevailed, allowing the
sun to beat down, heating the already warmer water. When seawater temperature rises, it holds less oxygen. Within weeks, the coral reef's fauna—
angelfish, surgeonfish, and butterflyfish—began to sicken. Their respiration
rates increased, and one report described the fish as "panting." Usually
skittish and agile species could be caught by hand. Corals turned white.
This was the first larger-than-local-scale bleaching event, and it was
connected to elevated water temperatures.

Two years later, the bleaching returned. This time, scientists connected it to the intermittent oceanographic condition known as El Niño,
which changes weather patterns around the hemisphere. This ocean-
sized disturbance was first noticed generations ago in Peru, where an
occasional warming of the seawater occurred around the Christmas season. The fishermen named it for the similarly timed arrival of "El Niño,"
the Baby Jesus. But the 1982 El Niño was no seasonal event. It lasted a

grueling two years. By February 1983, the coral reefs on the west coasts of Panama, Costa Rica, and Colombia and in the Galápagos Islands had bleached. Many of them did not recover: 70 to 95 percent of the corals to depths of sixty feet died. In the Caribbean, the heat stymied the northward currents bringing doldrums, similar to what had happened in 1980. As the water warmed, the corals turned white throughout the southern third of the Florida Keys. This was the first time the bleaching went global.

For the next decade or so, regional-scale bleaching reports popped up around the world with greater and greater frequency. In 1987 it was the Caribbean again along with the Maldives and Costa Rica; in 1991, Thailand and Tahiti; in 1993, Bali, Indonesia; in 1994, the Central Pacific including the Cook Islands and Guam.

All of that now seems like a low-grade fever compared to the pyrexia in 1997 and 1998, when the seas experienced the strongest El Niño up to that point. The corals responded with the greatest bleaching up to that point. Reefs bleached in at least thirty-two countries. In many places, it was the first time the corals had ever bleached. One of those countries was Australia. There, massive boulder corals, including one that was seven hundred years old—meaning it had survived all that the oceans had thrown at it for seven centuries—were killed by the heat. An assessment by the Global Coral Reef Monitoring Network showed that almost half of the coral reefs in the Indian Ocean bleached; one third in the Arabian Sea; and a fifth in South and East Asia. Worldwide, 11 percent of the reefs were lost and a further 16 percent remained damaged three years later. The global expanse of the bleaching was so great that scientists questioned whether what had seemed so shocking in 1982 and 1983 should still be considered a global-scale bleaching.

What was no longer in question was that elevated temperatures were a major reason for large-scale coral decimation worldwide, and that this was a new condition. Coral skeletons show growth rings like trees do: increased growth in the summer because of higher activity by their algal symbionts and slower growth with shorter days in the winter. In centuries-old coral skeletons sampled from Mexico, a new kind of mark was etched in the layers of calcium carbonate: stress bands, showing periods of mass bleaching.

Another El Niño hit in 2009. In 2010, reports of bleaching rolled in from across the Indian Ocean from Kenya to the Maldives. They blanketed Indonesia and Japan. They extended up and down the Caribbean from Curaçao and Venezuela to Barbados and Cuba. The mass bleaching encircled the globe yet again.

Terry Hughes, one of Australia's preeminent coral scientists, published work showing that as the ocean has absorbed more and more heat—remember those nuclear bombs worth of thermal energy deployed in the ocean every second?—background water temperatures have been climbing. Prior to 1980, El Niños raised the seawater temperature every four to seven years, but they never caused mass bleaching. In the past twenty years, the temperature increase from an El Niño that might have been tolerated before wasn't any longer. Hughes explained, "Inevitably, the link between El Niño as the predominant trigger of mass bleaching is diminishing as global warming continues and as summer temperature thresholds for bleaching are increasingly exceeded . . ." Now, bleaching occurs even without El Niño.

In 2015, another blast from the El Niño furnace torched the sea. Guam reported bleaching first. Then reefs on every island of the Hawaiian archipelago turned ghostly. Bleaching was reported throughout the eastern Pacific: Kiribati, Samoa, and as far east as Panama. Florida began bleaching in August, followed by Cuba, the Dominican Republic, Puerto Rico, and the Cayman Islands. The bleaching roiled west into the Indian Ocean and north into the Red Sea. By 2016, the warm water started a second pass, taking aim for the Seychelles and across the Indian Ocean to the Maldives, Mauritius, and India. Two thirds of the northern section of the Great Barrier Reef bleached. With barely a break, the heat returned in 2017, this time taking aim at a five-hundred-mile stretch of the Great Barrier Reef's central section. In Japan's Sekiseishoko Reef, 90 percent of the reef bleached. Vietnam, Thailand, Taiwan, Philippines, and Indonesia all reported severe bleaching. From Guam to Hawai`i, from Fiji to French Polynesia, corals underwent bleaching. This 2015–2017 catastrophe was what Richard Vevers had dubbed the Third Global Bleaching Event. It was the worst in history. He told *The Guardian,* "I can't even tell you how bad I smelt after the dive—the smell of millions of rotting animals." In total, three quarters of the reefs in the Indian Ocean and Pacific Ocean bleached. Bleaching was documented in more than 90 percent of the three thousand smaller reefs that make up the Great Barrier Reef.

IN 2017, Steve Coles repeated an experiment he'd performed in 1970 working on the heat outflow at the Hawaiian Electric Company plant. He exposed the same three species of Hawaiian coral to an increased temperature of 31°C. In 1970, after about a week at 31°C, all of the coral bleached. But in 2017, none of them did. In the forty-seven years between 1970 and 2017, the average water temperature in Hawai`i has increased 2.2°C. One explanation for the results was that the corals were already adapting to the increased water temperatures. Because the tools to study the genetics of the corals didn't exist in 1970, there's no way to tell if the populations have been shifting to individuals with genes for tolerating higher temperatures or if the corals were just getting tougher in the face of climate change. Regardless, "it was a real surprise," Steve told me by phone. "The big question was the rate at which it happened. It happened in under fifty years." That rate is critical. In order to outpace the expected rise in sea temperatures, corals need to increase their ability to handle heat by between 0.2°C and 1.0°C per decade. The corals in Steve's study increased their tolerance by 0.48°C per decade, squarely in the middle of that range.

When I first spoke to Misha over a cup of tea and chocolates, he mentioned a study suggesting that this kind of adaptation wasn't happening just in Hawai`i. "In the last decade the threshold for bleaching, heat tolerance, went up half a degree Celsius worldwide," he'd said. Looking at the probability of bleaching on more than thirty-three hundred coral reefs in eighty-one countries, researchers found that although bleaching events increased between 2002 and 2017, something else interesting happened at the same time: the temperature at which the coral bleached shifted. Two decades ago, the mean temperature of seawater that would bleach coral was 28.1°C. Today that temperature has increased to 28.7°C. The authors of the study wrote, "Past bleaching events may have culled the thermally susceptible individuals, resulting in a recent adjustment of the remaining coral population to higher thresholds of bleaching temperatures." The survivors, the ones that would live to reproduce, would be the ones that could handle the heat.

"Anecdotally," Misha added, "where I've worked in the Great Barrier Reef for the last ten years, it takes much more heating to make them bleach—like degrees Celsius. It seems to me that the corals are already much more badass."

There's another way to think about Steve's repeat study and Misha's anecdotal observation. Bleaching tolerance involves more than just the coral. It's a badass merger after all. While it's perhaps tempting for us, as part of the animal kingdom, to focus on the animal part of the story, the role of the algae bears some serious attention too.

In 1881, German biologist Karl Andreas Heinrich Brandt published a paper whose title translates roughly as "About the Cohabitation of Animals and Algae." (*Symbiosis* literally means the same thing. It's Greek: *sym* meaning "together" and *bio* for "living." The word was invented two years earlier, but its meaning didn't have the specificity Brandt was looking for, so cohabitation it was.) Brandt and a few other scientists working around that time noticed green- or yellow-colored cells in marine creatures like sea anemones, sponges, and jellyfish. After a fair bit of microscope work, Brandt determined that the pigmented cells were separate beings—algae—and coined the word *zooxanthellae* meaning, roughly, "yellowish little thing in animals." Research in the twentieth century placed the solar-symbionts in a group of single-celled algae called dinoflagellates, which are also responsible for red tides and the glittering bioluminescence sometimes seen in breaking waves. While they do prosper inside animal tissues, they can also be found swimming free in the sea, where their form is a single roundish cell with two long, tail-like flagella. One flagellum trails out the back, but the other wraps around a groove in the midsection. When it swims, it spins like a top.

When I first learned about zooxanthellae in college, I wondered why the Greek word for yellow was in their name if chlorophyll is green. But when I finally saw the algae tucked inside a coral cell, the reason was dazzlingly obvious. Zooxanthellae are golden and gemlike. They look like shimmering amber spheres embedded in the clear coral tissue.

That leads to another question: Why do algae look yellow? Zooxanthellae do contain green chlorophyll, but they manufacture other pigments too. These accessory pigments absorb much of the remaining blueish and greenish light, leaving only the yellow wavelengths to glimmer back toward our eyes. Collectively, accessory pigments boost the algae's productivity, like adding bandwidth to an antenna. The accessory pigments also stabilize the molecular machinery of the cell and mop up toxic spills of molecules, known as oxygen radicals, which can form during photosynthesis. Before I looked into it, I thought of zooxanthellae as something like static solar panels plunked on the roof of a house, the house being the coral. But that sorely underestimates zooxanthellae. These algae carefully control their solar power production by—among other processes—micromanaging accessory pigments depending on the temperature and light conditions. Bright, sunny day? Increase accessory pigment production because there are bound to be heat and oxygen radical spills. Cold snap? Downregulate. There's no need to waste energy on pigments that don't have a job to do. These tiny cells are fine-tuned solar systems.

Today, the word *zooxanthellae* is used less frequently among coral scientists. They usually call them by their family name, Symbiodiniaceae. Teasing apart the genetics of the Symbiodiniaceae has been complicated, and there are thought to be hundreds of undescribed species, each with particular host animals, temperature preferences, and geographical distributions. One reason for the complexity is that the genome of these symbiotic algae is huge. They have a whopping fifty-nine thousand genes—about three times more than we humans have—and their DNA is stored in a peculiar kind of chromosome that's hard to sequence. For years, scientists just called the different varieties of zooxanthellae "clades," using consecutive alphabetical letters as new ones were described. The naming had reached Clade V in 2018, when a major taxonomic revision of the group was finally performed. One outcome of that analysis put the evolutionary origin of zooxanthellae at 160 million years ago, just around the time the number of modern coral species was burgeoning, implying that the algal symbionts were critical to coral diversification. The other important result of study was that it slimmed down the twenty-two lettered clades of zooxanthellae into seven official genera, with actual names rather than just an initial.

What's turning out to be most important about all these different groups is that they might represent options for a coral in a future warmer ocean, like having a choice of power company. Where I live in Austin, you can choose wind power over fossil fuel power, so you might want to give wind a try to keep your carbon footprint down. On the other hand, wind power costs a bit more, so if your budget's tight, it might not be feasible. In the same way, each group of algae has slightly different abilities to absorb different wavelengths of light, to produce sugar, and, importantly, to withstand different temperatures. Members of Clade C, now known as the genus *Cladocopium*, are quite accommodating. They share almost all of the sugar they make through photosynthesis with the coral. But they are sensitive too. They don't handle changes in temperature well and are therefore prone to bleaching. Members of Clade D, or *Durisdinium*, drive a harder deal. They keep more of their sugar for themselves, skimping on passing it to the coral. But *Durisdinium* have a thermal threshold as much as 1.5°C higher than *Cladocopium* before bleaching.

Most coral species can host different clades of algae, and scientists have started to look at changes in the populations of algae with temperature shifts and in the wake of bleaching. In Australia, scientists transplanted corals from cooler waters in the southern part of the Great Barrier Reef to warmer waters in the north. Originally, the coral were dominated by *Cladocopium*, but also housed a small population of *Durisdinium*. Upon reaching the warmer water, the coral immediately bleached, but they didn't die. Instead, they recolonized entirely with *Durisdinium*. In American Samoa, when El Niño brought uncharacteristically warm water to the reef in 2015 and 2017, scientists documented corals switching from *Cladocopium* to *Durisdinium*. The coral colonies that switched bleached less, but they also grew more slowly after the heat wave.

The idea that coral might shuffle symbionts to survive the warmer future is still under study, but it's very intriguing. Whether or not the shuffling will enhance survival given that an alga like *Durisdinium* is a more demanding, selfish power source is still unknown. Whether corals can acquire *Durisdinium* quickly enough to make a difference is also undetermined. The authors of the study in Australia called the possibility of symbiont shuffling a "nugget of hope." They wrote, "While this is likely to be of huge ecological benefit, it may not be enough to help these populations cope with the predicted increases in average tropical sea temperatures over the next 100 years. . . . It may, however, be enough to 'buy time' while measures are put in place to reduce greenhouse gas emissions." Like all relationships, getting along is a kind of dance that both partners are constantly evaluating. While we above the seas are dithering in addressing climate change, out on the reefs it might be switching partners that makes all the difference.

10

Unmasking Immunity

The Society of Integrative and Comparative Biology goes by the acronym SICB, pronounced "sick bee." And I'd bet that in any year, the society's annual meeting is certain to have several talks if not an entire session on apiary disease. The 2020 annual meeting included symposia called "Biology at the Cusp," looking at reptile teeth; "Fire When Ready," about muscle activity; and "You Got the Touch (and Sight and Sound)," looking at animals' ability to sense the world. Lucky for me, the meeting also featured a full day of talks under the heading "Beat It (the Heat): Combating Stress in Coral Reefs." And, conveniently, the meeting was held in downtown Austin.

I arrived just a few minutes early and found a seat toward the center of the room. I gazed around, trying to see if I recognized anyone. At the same time, I did the kind of quick count I always do at science meetings. Women outnumbered men about 2.5 to 1. It is the kind of ratio I have become used to seeing at marine science conferences, and I still don't know whether it's the field that attracts more women or whether women are more likely to come to conferences like these. A recent study by the National Science Foundation showed that in the past decade, women earned half the marine science doctorates. But at higher levels of science, such parity is still evolving. In 2021, for the first time, half the inductees into the National Academy of Sciences were women. But between 1994 and 2011, the number of men elected remained about six times that of women.

As I swiveled back around in my seat, I caught the eye of an older man walking in the door. Recognition dawned on both of us, but neither of us could put a name to the face. And then something brought him into focus for me. He'd been a professor in my department when I was a graduate student. And he had mocked my research, deriding it for being too math-heavy for a biology department. The acrimony got stupid when I went to the department office to pick up a laser pointer on the morning of my PhD defense, a talk that was to be the culmination of all the work I'd done over the previous six years. The secretary said this professor had just come to get them all to change the batteries. That left me running to the school bookstore in my high heels to buy a laser pointer in the minutes before the most important talk of my life. I don't like admitting that I still hold a grudge.

The man sat down next to me, and we shook hands politely. He said he'd heard I was now a science writer and that throughout his career he'd always encouraged good science communication.

I confirmed that yes, I was a science writer, and pondered if he was complimenting himself or my career choice. Then I blurted, "What brings you here?" I regretted it instantly. This was exactly the kind of meeting that suited his research.

As he talked about himself and his research, something he'd always done well, I felt my chest unwind. Someone who had once been a towering figure was no longer intimidating. I had followed my path and he had followed his. It was unexpected and a little uncomfortable to be sitting next to him more than two decades later. But that was all. After a few presentations on coral, he handed me his card embossed with an impressive seal. I handed him mine, one I'd printed myself off my home computer. And he left.

A few rows in front of me, I noticed a hexagonal mandala tattoo in the center of the back of someone's neck. That could be only one person. During a short break, I scooted down the aisle to say hi to Misha Matz.

He invited me to join his lab and some other coral scientists for a beer after the meeting.

"I'd love to," I answered, and then ducked back to my seat.

The next talk was by one of Misha's students, Evelyn Abbott. I sat up a little straighter when she began by posing the same question that Maurice Yonge had asked ninety years ago in Australia's Low Isles: Who starts the process of bleaching, the coral or the algae?

I had been trying to answer this question myself for months. When I'd asked Misha for help understanding it, he'd suggested I call Virginia Weis at Oregon State University. Virginia has spent much of her career studying the symbiosis at a cellular level and confirmed that the intricacies of the breakup were still at the center of intense research, but that some progress had been made. She said that bleaching appeared to be connected to the coral's immune system. Immune systems are in the business of recognizing and attacking foreign particles that make their way inside the body. But in the case of the badass merger, when everything was functioning normally, rather than launch an immune response against something foreign—the algal cells—living inside its own tissues, the coral's immune system ignored their presence. How and why were unclear. The leading hypothesis was that the immune system ignored the algae because the algae biochemically disguised themselves from the immune system. But it could also be that the coral's immune system recognized the algae and sent out a message not to attack. Regardless, Virginia said, "bleaching is a waking up of the immune response. It seems as if the algae takes off its invisibility cloak, allowing the host to mount an immune response. But we don't have all the arrows. In fact, we still don't know who throws the first switch."

Evelyn's hypothesis was that the algae, not the coral, flicked the switch starting the bleaching process. Her reasoning was an evolutionary one. "They escape from a sinking ship," she said. By leaving the coral, the algae can swim away and live just fine—albeit without the easy supply of nitrogen and carbon dioxide they get when they are inside the coral. The algae can manage out in the seawater. The coral, on the other hand, can't survive very long without the fuel from the algae's sugar. Bleaching benefits the algae but not the coral.

Evelyn, Misha, and one of his postdocs, Groves Dixon, had performed an experiment they called "bleaching every which way" in which they bleached coral with heat, with cold, with low salinity water, and with mixtures of all of them. In each case, they measured changes in the thousands of different genes that were turned on and off in both the coral and the algae. The results were displayed on what Evelyn called "volcano graphs," which looked like the trace of lava erupting from the mouths of two volcanoes. A bigger explosion of lava indicated more activity happening inside the cells. In every case, the coral's lava bursts were massive compared to the algae's. What that meant was that during bleaching, the coral was a flurry of gene activity while the algae were barely registering any changes at all. "The genetic response from the coral was much greater than that of the algae," Evelyn concluded.

After her talk, I raised my hand to ask a clarifying question. "Circling back to your original question, what does the fact that the corals are launching such a huge genetic response and the algae aren't mean in terms of who starts the bleaching?"

"It indicates that my hypothesis was wrong," she answered, "and the coral starts it."

LATER THAT EVENING, the coral scientists from the meeting gathered at a bar in a hip part of downtown Austin. I ordered my favorite local beer, Electric Jellyfish, and not just because of the name, though it always brings me joy to ask the bartender for it. Walking into the crowd, I found myself in a conversation with Misha and Evelyn.

"So, do you really think the coral starts the bleaching?" I asked.

"I don't know," Evelyn said. "The data show the coral doing a lot. And it seems to be confirmed by other studies."

I nodded. Earlier in the day, I'd been to a couple other talks that described the coral's genes very busily handing out instructions in response to stress, while the algae plinked along like nothing was happening.

Sarah Davies, a professor at Boston University who previously worked in Misha's lab, joined us. She suggested a twist on the usual story that bleaching is bad news for the coral. "What if bleaching actually benefits the coral?" she asked. Sarah brought up a study suggesting that bleaching was all part of coral's immune system ramping up. Sure the symbiosis fell apart, but with a ramped-up immune system, perhaps the coral was better able to fight off dangerous pathogens.

Misha shook his head. "I still think it's the algae that start it," he said. "The algae don't have anything to lose, so evolutionarily it makes sense. It might even benefit the algae's fitness. You know, get out in the sea, have some sex. Mix up the genome."

He had a good point. Whenever a population could mix its genes, it created more variability and that meant more ability to withstand stress. Stuck inside a coral, the algae could clone only themselves.

"But what about Evelyn's data?" I asked.

"Uncloaking could be a small signal," Misha answered. "It might not take that much to show the coral who you are and create a big response in the coral."

The question of who starts the bleaching and exactly how it is initiated spun around and around in my mind as I sipped my beer and the coral conversations around me grew louder and more enthusiastic. Earlier in the day, I had sat next to a man who, through a very small action, had provoked an enormous response in me. Those actions left a lasting anger that spanned decades. Even so, today I smiled and was polite, as was he. I flaunted my professional success to him. He did the same to me, even more so. But I also knew that he'd had real difficulties in his personal life. Yet I didn't tell him that my daughter was suffering from a mental illness we had so far been unable to treat, that was destroying so much in her life and our family's. The relationship between coral and algae is every bit as complicated and sophisticated as the relationships among us humans. We share bits of ourselves and hold back others. We provoke responses. We react in ways that are sometimes disproportionate. It's a constant calculation, a constant conversation. We are always deciding what to reveal and what to hide.

WHEN ISY WAS VERY YOUNG, I remember browsing a Lucite holder full of paper brochures in the waiting room of a doctor's office. Out of boredom, I plucked one from its slot. PANDAS, it said in vertical letters top to bottom, with a small black and white panda bear tucked into the right corner. Intrigued by the odd name, I flipped it open and read about a condition called pediatric autoimmune neuropsychiatric disorders associated with Streptococcus infections. The details were fuzzy, but I remembered that symptoms included an abrupt onset of OCD following an infection by the streptococcus bacterium, one that is much more infamous for causing sore throats. This unusual disease might have stuck with me both because of the quirky name and because I'd never heard of a link between a bacterial infection and a psychological disorder. As we struggled to diagnose Isy's symptoms, PANDAS popped back into my mind. Googling the list of indications, ten of the eleven seemed to point to Isy. Was strep the underlying cause of her OCD? Over winter break in eighth grade, just before she started isolating herself from friends and failing her schoolwork, she'd had a strep exposure. It wasn't much. But it was a coincidence I couldn't shake.

I discovered that *Streptococcus pyogenes* (or Group A Strep) is capable of so much more than causing a sore throat. Strep has learned to disguise itself within our bodies, masquerading from our immune system with biochemical camouflage similar to the exteriors of our own cells. When our immune systems make antibodies to flag strep as foreign, those antibodies can mistakenly stick to our own cells. This "cross-reaction" is known to cause rheumatic fever. The antibodies lodge in the joints, kidney, and heart and cause painful inflammation there. In patients with PANDAS, the theory went, the cross-reacting antibodies were shepherded through the blood-brain barrier via nerves in the back of the nose. Once in the brain, they targeted the cells of basal ganglia, the ancient organ responsible for fight or flight. This awakened an immune response, an inflammation that caused anxiety and fear. Was it possible, I wondered, thinking back to the day Isy screamed, "Get it off of me!" and the day she'd looked blankly at me trying to remember the parts of an atom, that something had attacked her brain? As scientists suspect is the case with coral bleaching, were Isy's problems caused by her own immune system?

11

The Holobiont

I asked Isy's therapist and psychiatrist about the possibility that she had PANDAS, and they were both initially hesitant. There is not a simple test for PANDAS, and many in the medical community are skeptical about it. But, as Isy's symptoms intensified, they both agreed that we should look into it. The next step was a visit to an infectious disease specialist. Elevated levels of strep antibodies can indicate an unresolved strep infection. When her blood tests came back, those antibodies were double what's expected for someone with an active strep infection, an indication of PANDAS, though not proof. Some people just maintain higher antibodies following an infection. To test the PANDAS hypothesis, Isy was prescribed a heavy course of antibiotics. If strep was hiding in her body, the antibiotics would clear the bacteria, and her immune system would respond by reducing the production of cross-reacting antibodies. Maybe, I hoped desperately, Isy's problems could be solved not by extensive therapy and antianxiety drugs but by plain old antibiotics.

"Make sure she gets probiotics too," the doctor admonished. The doctor had put her on an antibiotic capable of killing many different kinds of bacteria, not just the *Streptococcus* we were trying to eradicate. The word *probiotics* is a mash-up of the Latin "for" and the Greek "life." Probiotics contain live yeasts and bacteria thought to help rebalance the community of microorganisms that normally live in the digestive system. Whenever

I doled out probiotic gummy chews along with probiotic fortified yogurt, I had the urge to clink my coffee cup with her juice glass and break out into a cockeyed version of the old *Fiddler on the Roof* song "To Life." Our microbiome is credited with supporting the health of many other organs, including hearts and livers; slowing the effects of aging; and aiding immune systems. That last one—probiotics supporting immunity—might be key to coral health too.

Corals have a huge surface area for their size. They are basically a thin veneer of living tissue spread over a stony skeleton, like melted cheese on top of macaroni. And like melted cheese, that tissue is the tasty part. So corals have to be capable of fending off hordes of would-be attackers. As a result, they set all variety of barriers and traps. One way corals protect themselves is by maintaining an artillery of stinging cells, called nematocysts. Just like the stinging cells in jellyfish, nematocysts are tiny capsules loaded with long poison darts. When triggered, the darts are launched at an acceleration five million times that of gravity. Take a pen and drop it on the table—that's the acceleration of gravity, 1g. Try to imagine a motion five million times faster—*that's* how fast a coral's stinging cell deploys. Stinging cells are complex, sophisticated weaponry, effective and powerful.

Like the insides of our noses, corals also use a barrier of mucus that acts like a gooey biochemical moat. Some corals produce so much mucus that when they're held in the air, the goo drips off in sheets. Underwater, mucus forms strings and webs that barricade the surface of the coral. Studies from the 1980s suggest that as much as half a coral's energy goes into making mucus, which says a lot about how important it is to survival. Mucus is a mixture of sugars and proteins, basically bacteria food. And lots of bacteria grow there: somewhere between five million and five hundred million per teaspoon of goo. Some of the bacteria attracted to the coral's mucus seem to help fight off pathogens, like the one that causes a disease called white pox. And like humans blowing their noses, corals regularly shed their mucus layer, washing away accumulated sand

and sediments. Rather than tossing away all that accumulated organic matter, corals sometimes eat their mucus, making a snack of everything growing in it. (It's okay, go ahead and make a joke about coral eating their boogers.)

Just like in us, bacteria line the guts of coral, and they seem to serve some similar roles as in our guts. There's a bacterium in coral mucus that makes vitamin B_{12}, a chemical critical for all animals, but which animals can't produce. Likewise, we humans harbor bacteria in our own guts that make vitamin B_{12}. And just as gut microbiomes are unique to each of us, different coral species have unique microbiomes. Corals of the same species living thousands of miles apart were found to harbor similar bacteria communities that were unlike those found in other species nearby. And like us, microbiomes shift when corals are stressed. Populations of *Vibrio* bacteria (relatives of the bacterium that causes cholera in humans) increase when a coral bleaches but return to normal levels after a coral recovers. One species of coral shifted its microbiome when transplanted from clean water to more polluted surroundings. It then shifted back when returned to the healthy environment, recovering its original bacterial entourage. Corals also shift their microbial communities in the summer and the winter, something that's been observed in humans too: bacteria that cause bad breath and flatulence peak in January, those that cause foot odor in February, and those that cause underarm odor in July.

As awareness of the dynamic world of microorganisms associated with corals grew, in 2006 scientists proposed the coral probiotic hypothesis. It suggested not only that algae are critical to the health of coral, but that bacteria and other microbes are as well. Similar to the solar shuffling of the algae, switching and selecting bacteria that live with coral might provide coral with a quick way to boost their immunity. The researchers wrote, "Changing their microbial partners would allow the corals to adapt to changing environmental conditions more rapidly (days to weeks) than via mutation and selection (many years)." In other words, rather than waiting for evolution to mutate the right genes to make the right proteins to fight a disease and then for the plodding process of spreading those genes throughout a population, recruiting bacteria that are already equipped to fight the pathogen could be like getting a quick shot to the immune system.

There's some evidence that probiotic jockeying might occur in the wild. In 1995, researchers from Israel noticed that a species of coral from the Mediterranean called *Oculina patagonica* had begun bleaching. All of the bleached colonies were infected with the pathogenic bacterium *Vibrio shiloi* (another relative of the cholera bacterium), but none of the healthy ones showed any sign of infection. Five years later, the bleaching occurred again. Again, the researchers found *Vibrio shiloi* in huge abundances in all the sickened colonies. This trend continued in 2001 and 2002: bleached colonies were always infected with *Vibrio shiloi*. But in 2003, the researchers tried to infect newly collected coral with cultures of *Vibrio shiloi*. Nothing happened. They wondered if their bacteria cultures had developed some sort of mutation that prevented them from infecting coral and ordered a fresh batch. Again, the coral couldn't be infected and didn't bleach. Looking closer, the researchers saw *Vibrio shiloi* were attaching to the corals and penetrated the coral tissues. But instead of multiplying there, the bacteria started dying. The corals seemed to have developed some sort of immunity to the pathogen. But how?

Animals have two types of immune systems: innate and adaptive. The innate immune system is a general one that besieges any pathogen. It causes inflammation, which recruits cells that can attack all kinds of invaders. It's not specific. The adaptive immune system targets a specific pathogen, as the antibodies we make do to flag a specific virus as problematic. Unlike us, corals have no adaptive immune system. They can't make new antibodies when exposed to new pathogens. The researchers believed that the newly acquired immunity to a specific pathogen came not from the coral itself but from another microbe. They wrote, "Corals lack the ability to produce antibodies and have no adaptive immune system. The Coral Probiotic Hypothesis would predict that one of the properties of the bacterial community which is currently established in *Oculina patagonica* is the ability to produce materials which bring about the lysis [the technical term for cell rupture] of *Vibrio shiloi*." *Oculina patagonica*, they suggested, had gotten a probiotic vaccine.

In the 1990s, lauded evolutionary biologist Lynn Margulis introduced a theory for the origin of complex cells with organelles like those in our bodies and the bodies of plants and fungi. She hypothesized that long ago, two cells teamed up. One cell became subsumed in the other, lost its ability to live independently, and took on the job of gene storage in the case of a nucleus or energy production in the case of mitochondria or solar power in the case of the chloroplast. Even the smallest unit of life—cells—are not stand-alone things but cooperative consortia. Margulis called this collective unit a holobiont, *holo* meaning "whole." The scientists from Israel studying the newly immune *Oculina patagonica* extended this term *holobiont* to emphasize that the coral is much more than just the coral animal. It is even more than its animal part merged together with its solar-powered algal partner. A coral is a multi-kingdom conglomerate, comprising animal, plant, and bacteria, and probably fungi too. Of course, all of us are more than just our animal part. We all contain members of other microscopic kingdoms: some beneficial, some parasitic, and some whose role resides somewhere along that spectrum. We are all compendia.

But there is something so fundamental about the way that corals gather others around and inside themselves. Without energy from algae and without a network of mucus harboring a complex community of bacteria to support its health, corals would be incapable of existing. The word *coral* falls well short as a descriptor of the inclusive, complex nature of the assemblage. Today, at most coral meetings and in many coral publications, I often hear the scientists who work in the field refer not to the animal but to the holobiont, the incorporation of the coral colony and all its many affiliates. While it is easy for us to forget that we are supported by our microbiome, it is impossible to talk about coral without acknowledging that it is much greater than its individual parts.

A FEW DAYS AFTER she began the antibiotics, I awoke and didn't see Isy standing above me, hair astray, peering at the clock on my nightstand for the time. Her OCD, like so many other denials, refused her the luxury of having a clock in her room. But on this day, just the dampened sunlight streamed through the curtains. I'd woken to her sleep-smudged face so many days, it had become its own sort of familiar alarm clock. Its absence felt like I'd overslept for an important meeting.

I scooted quietly out of bed and around the corner to the den. There, encased in the morning light, was my daughter, curled in a ball with a red plaid blanket around her knees, a dog tucked between her feet and the coffee table. In her hands, she held a book. Her face, relaxed and unanxious, was gently engrossed in the words on the page. Just a week earlier, even reading a sentence on her computer was impossible. Seeing her with a book in her hands was a sight I hadn't glimpsed in over a year. I felt a surge upward from that place in our hearts that holds a sacred song for moments like these. I didn't know if the antibiotic was the answer to her OCD. I didn't know how to contain my expectations, to neutralize my optimism. I told myself to try to walk the line of objectivity. But I also knew that as a mother, I couldn't suppress my feelings of hope.

12

Throwing Shade

The theater where I saw *Chasing Coral*—the movie starring The Ocean Agency's founder Richard Vevers—was part of a local chain where you can order dinner and a beer along with your show. I was running late, so I slunk in the back and took a seat, scribbling my order for a pizza and an Electric Jellyfish on a scrap of paper for the waiter. The screening was sponsored by a nonprofit whose mission is to encourage kids to connect with nature. Before the movie started, the group's executive director introduced a high school student with long blond curls who stood shyly and waved. She had made a short film about coral, which would show before the feature. The lights dimmed and the footage opened to a boat's wake, white foam against the deep blue. The camera plunged underwater, panning over Florida sea fans and schools of yellow and teal-striped fish. Words flashed on the screen: "Over 50% of the coral reefs have died in the last 30 years. What if there was a way every person on Earth could help?"

The movie shifted to interviews with coral experts who laid out the basics of bleaching. At that point, I expected the film to highlight the problem of warming seas. Instead, "In the 1970s, oxybenzone was introduced in sunscreens in order to help us play in the sun. But unfortunately, it has been found to have very significant effects on corals and animals that live in the tropical marine environment." The speaker created a kind of dome-like shape with her hands. "Specifically, when a coral larva is in the presence of oxybenzone, the skeleton actually grows around the larva itself. Imagine your skeleton growing out through your arms and entombing you in a tomb of your own making, basically burying the coral alive."

That was unexpected.

Text flashed on the screen again. "One drop of oxybenzone in six and a half pools of water can harm coral," followed by a list of four other chemicals also presumed dangerous to corals: octinoxate, homosalate, octisalate, and octocrylene. These chemicals are thought to be harmful in very small amounts, but that makes sense. They work like hormones, triggers that set off a larger cascade. A tiny bit goes a long way.

Although synthesized from fossil fuels today, German biochemists discovered oxybenzone in plants more than a century ago. Like us, plants probably use it as a protection against the harmful rays of the sun. Oxybenzone consists of a six-sided ring of carbon called an aromatic. The chemical bonds in that ring absorb ultraviolet light, especially in the part of the spectrum known as UVB, which causes sunburn and wrinkles. Beginning in the 1940s, awareness of aging effects of the sun's rays pushed makers of personal care products to add sunscreens into their lotions and cosmetics.

The two most cited scientific articles on sunscreen and corals were published in 2014 and 2016. Craig Downs, a coral scientist at the non-profit Haereticus Environmental Laboratory, was lead author on both of them. Craig discovered the problem while diving in Trunk Bay in the U.S. Virgin Islands National Park. There, he encountered a large stand of dead elkhorn coral, their skeletons flipped over by the waves. Upside down, the splayed branches reminded him of wings and the stalk a weapon. "It looked like something from the Book of Revelation," Craig said, "like the war of the angels, and all the angels had been impaled on spears." After that dive, Craig and his research team discussed the devastation as they stopped in a small grocery store. A local man overheard them and said, "It's the tourists." The beach adjacent to the dead corals often hosted as many as five thousand visitors a day. The local man suggested Craig return at sunset, which he did. The water's surface glistened in an iridescent sheen of sunscreen.

Back in his lab in Virginia and later working with scientists from Israel, Craig tested the effects of the chemicals in sunscreen, especially oxybenzone, on coral larvae. At 23 micrograms per liter (about equivalent to the mass of ten raisins in an Olympic swimming pool), they stopped swimming. At tenfold higher concentration, the larvae began to bleach. If the concentration was increased a hundredfold, the larvae became opaque from overproducing calcium carbonate in their skeletons. That's when they encased themselves in a tomb of their own making.

But questions linger. Craig used a species of coral larvae (*Stylophora pistillata*) that is also among the minority of corals that brood their larvae internally rather than spawn into the open sea. Brooders produce larvae year-round, making them useful for lab work, but not always representative of more typical broadcast larvae. And the species he used lives in the Pacific. To this day, no one has tested how oxybenzone affects coral larvae native to the Caribbean. Replicate work on other species was finally published in 2019. At concentrations forty times higher than in Craig's work, one species experienced 5 percent mortality, but larvae from another species were not affected at any concentration of oxybenzone tested. Another 2019 study exposed two species of adult coral to both oxybenzone and elevated water temperature. That paper was titled "Adding Insult to Injury." The takeaway was that the fatal blow came from warming, while sunscreen was merely an added insult.

And there's another important question: Are concentrations of oxybenzone in the ocean high enough to be problematic? In a 2018 review of two dozen studies by the International Coral Reef Initiative, just one, from Trunk Bay in the U.S. Virgin Islands, reached concerning levels at 1,395 micrograms per liter. But measurements of oxybenzone from Hawai`i, even places where five hundred or more swimmers were in the water, were barely detectable. In the other studies, from Hong Kong, Palau, and Japan, among others, oxybenzone could not be measured.

The question of what chemicals in sunscreens do to humans is just as confusing as it is for coral. In 2001, the first alarm bells began ringing, with oxybenzone sounding the loudest. A study that year showed that oxybenzone could mimic estrogen, increasing uterus size in immature rats by about a quarter. In the years since, oxybenzone has been linked to a number of human health issues, especially related to reproduction, including endometriosis and low sperm viability. It has been found in breast milk and implicated in birth defects. However, a 2017 review of the health impacts of oxybenzone questioned the methods of some of these studies.

While the preponderance of oxybenzone measurements out in the sea have been low or undetectable—at least so far—the same may not be true for our own bodies. The Centers for Disease Control and Prevention studied 2,517 urine samples from people aged six and older. They found oxybenzone in nearly everyone: 96.8 percent of the samples. It wasn't clear that the source of that oxybenzone was sunscreen alone. Oxybenzone is also found in plastics, fabrics, and outdoor furniture—things that we want to protect from the damaging effects of the sun—so it could be absorbed into our bodies from other places, and it might be much more absorbable than previously thought. In a 2020 randomized clinical study of forty-eight individuals who applied sunscreen on 75 percent of their bodies four times a day, oxybenzone blood concentrations exceeded FDA limits with just one application and maxed out fifty times higher. The result triggered a requirement for more testing. The FDA recategorized oxybenzone and eleven other chemicals, removing them from a category known as GRASE (generally recognized as safe and effective) and requiring additional evaluation, which was under way as of 2020.

Health advocates and industry advocates have been lobbying for the chemicals in sunscreens to get a second look for decades. In Europe, active ingredients in sunscreens are classified as cosmetics, while in the United States they are regulated as drugs, requiring long-term and expensive studies to ensure safety. This bottleneck has stymied innovation in sunscreens. In the United States only sixteen UV-blocking chemicals are approved for use, and functionally only eight of them are used. But Europe has twenty-seven on the market. Some of those can be used at lower concentrations, making sunscreens feel less goopy and potentially adding fewer chemicals to the environment. All twenty-seven of the sunscreens approved for use in Europe have been submitted to the FDA for approval, but no progress has been made for decades.

Corals themselves may hold some answers to these problems. Marine creatures are exposed to the sun's damaging UV rays too. They manufacture their own sunscreen, molecules called mycosporine amino acids, or MAAs. More than twenty MAAs are known, and they accumulate in tissues that have the most to lose from exposure to the sun: the skin of sea cucumbers, floating sea urchin eggs, the lenses of fish eyes. Algae are thought to produce MAAs, and animals acquire them through their diets. Corals, of course, have that badass relationship with algae, and simply trade nutrients for sunscreen. MAAs are particularly good at absorbing the part of the sun's spectrum known as UVA that is most implicated in DNA damage and melanoma. MAAs may also have antioxidant properties. Just a few studies have looked at MAAs and human cells—all of them in culture—but the results are promising. One even suggested that MAAs activate a slew of genes involved with DNA repair and inflammation. But without the economic promise of FDA approval, the potential for MAAs in sunscreen remains unexplored.

In an email response to my questions about oxybenzone, a representative from the FDA wrote that the Coronavirus Aid, Relief, and Economic Security Act, or CARES Act, implemented at the beginning of the pandemic, gave the FDA the ability to speed up the process for reviewing chemicals like those in sunscreen. It also incentivized industry to explore bringing new products to market. Unexpectedly, as a result of a virus, some of the confusion and bottlenecks that have dogged sunscreens for decades may start to clear.

In the meantime, some governments have begun taking steps to ban the sale of sunscreens containing oxybenzone and other chemicals. National parks in Mexico have begun asking visitors to avoid chemical sunscreens. In 2018, the island of Bonaire, the state of Hawai`i, Leleuvia

Island in Fiji, parts of Mexico, and the country of Palau all banned oxybenzone. A few months later, a ban was passed in the Florida Keys, although it was contested in court. In July 2019, oxybenzone and octinoxate were banned in the U.S. Virgin Islands. In October 2020, in Sonoma County, California, the DA successfully prosecuted a case against a sunscreen company for falsely marketing "reef safe" products. The sunscreen company agreed to pay $50,000 in restitution, some of which benefited the National Fish and Wildlife Foundation. The press release on the judgment stated, "There is consensus among the scientific community that ingredients in sunscreen harm coral reefs." But the more I looked into it, the less certain I was that consensus exists.

Trying to sort out the conflicting evidence, I took every chance I could to ask coral scientists about coral and sunscreen. I put out a call on coral Twitter asking for input. While certainly not a comprehensive study, all except one said they didn't believe that sunscreen was a major problem for coral, especially compared to problems of illegal fishing, wastewater management, and the most critical, climate change. Misha said his lab had performed unpublished experiments exposing corals to sunscreen and found little to no impact. "It's possible urine is more of a problem."

Why, then, would so many people become worried about sunscreen and so many governments move to ban them? I asked. One answer was that it's relatively easy to do, much easier than tackling those bigger problems. You've heard of greenwashing? one scientist asked, referring to the act of making the appearance of sustainability while continuing to contribute to environmental degradation.

One of the scientists I polled by email was Megan Morikawa, who had given a talk about the role of the hospitality industry in coral reef restoration at the Reef Futures meeting. She worked with the Spanish hotel and resort company, Iberostar, managing their coral restoration program. Megan acknowledged how complicated the topic was. "We decided to invest in only sunscreen that did not contain the chemicals [oxybenzone and octinoxate] at Iberostar. Even though," she wrote, "what we did with the coral lab and our in-water operations was quantifiably far more helpful to our corals in our reefs. However the 'stickiness' of the topic, despite limited scientific evidence, is perhaps an example of how citizens want to do something to help reefs but need clear, concrete actions that they can take with their purchasing power to be able to do so."

Megan was right. We do need to feel like our actions matter. The student film about sunscreen I had seen before *Chasing Coral* urged beachgoers not to forgo sun protection but to choose clothing instead whenever possible. If it was necessary to wear sunscreen, the film suggested choosing brands containing non-nano mineral sunscreens like zinc oxide and titanium oxide instead. And although it remains unclear what it means to the health of either coral or ourselves, I support those recommendations. I got up from my computer and gathered up all the sunscreen bottles that had accumulated around our house. I inspected each one for oxybenzone and any other potentially problematic ingredients. Sealing them all in a plastic bag to prevent leakage, I dumped them in the garbage.

PART III

FLORIDA

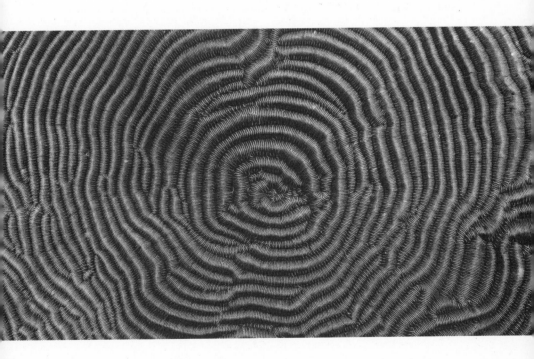

13

Fragmentation

round two hundred million years ago, the ancient Tethys Sea opened a rift that split the continents to the north and south. On the land's far west edge, a nook formed. It was in the shallow waters of that nook that the Caribbean was birthed, and its corals began to build their reefs. A quarter million years ago, the skeletons of those corals encircled the sea's hundreds of islands, forming jungle-like underwater thickets. The dominant species are known as elkhorn (*Acropora palmata*) and staghorn (*Acropora cervicornis*). In Latin, the word *palmata* is a large lobe or palm like an elk's wide antler. The *cervi* refers to deer and *cornis* to horned. The reefs these two species built, along with about three dozen of their more boulder-shaped kin, formed the backbone of massive living limestone metropolises stretching more than six hundred miles along the Yucatán and into Belize, and over three hundred more from north of Miami down to the tips of the Keys, the second- and third-longest barrier reefs in the world, respectively.

For more than 250,000 years, the sunlit Caribbean waters remained a vibrant tangle of skeleton, polyp, and tentacle and all of the great diversity of life sustained within. But around 1950, the coral thickets began to thin dramatically. For reasons that are unclear—rigorous ecological studies began only after the fact in the 1960s—the dominant *Acropora* numbers fell by half. One proposal suggests that fertilizer and pesticide runoff from rapidly agriculturalizing economies degraded the water chemistry, weakening the coral. In the 1970s and 1980s, Caribbean corals were attacked by an affliction called white band disease. This epidemic killed between 80 and 98 percent of the remaining staghorn and elkhorn colonies. In the following decade, another pathogen ravaged the region. This one targeted one of the corals' major allies: a lime-sized sea urchin with wicked black spines called *Diadema antillarum*. These voracious herbivores played a key role by clearing the reefs of prolific turf algae that can edge out coral for territory. Ninety percent of the sea urchins died. Other vegetarians, like the gaudy-colored parrotfishes that might have helped keep the seaweed at bay, were already overfished. An explosion of leafy and mossy algae ensued, one that has never really been quashed. And then there have been the storms. While coral in the Caribbean tolerated hurricanes for hundreds of thousands of years—they are a natural occurrence after all—recently the corals have increasingly failed to recover in their wake. All of that damage was the setting for the localized bleaching events of the 1970s, which became regional in the 1980s, and continue today. In the most severely affected parts of Florida, as little as 2 percent of the original skeleton-building coral remains alive.

AROUND THE TIME I went to the Reef Futures meeting, a video came across my Facebook feed called "Saving Coral Reefs One Coral at a Time." It featured a man named Dave Vaughan with a long white beard and unruly hair—he looked like a cross between Santa Claus and Jimmy Buffett. The video was put out by AARP and included clips of Dave scuba diving among broken forests of coral, swimming among healthy coral nurseries, and declaring that he wouldn't retire until he personally planted a million colonies back on the reef.

Dave's journey to making this unusual vow began with a career in aquaculture, working to sustainably raise clams, oysters, and fish for the aquarium trade. These endeavors grew into helping to build one of the largest ornamental fish companies in North America, on the campus of Florida Atlantic University. As the business became more successful,

Dave wondered if he might try growing corals for the aquarium trade. He set up a few tanks and was soon producing thousands. One day, Jacques Cousteau's grandchildren Philippe and Alexandra came to tour the facility. Dave pointed out the new coral he was planning to sell into the aquarium industry. He expected the Cousteaus to be impressed, but Philippe shook his head. "Dave, you've missed the point. If you can do this for the aquarium trade, you can be doing this for the reef."

In a YouTube interview, Dave explained why the idea was revelatory. Restoration was common on land, but not undersea. And there was a difference that is so simple it almost seems unnecessary to mention, yet it is vast. On land, we've long known how to grow trees and grasses. When it comes to the seas, we are just beginning. Although in recent decades we've begun to farm fish and seaweed at scale, coral was never considered a crop. But Philippe's words inspired Dave to try. He switched jobs, becoming senior scientist at Mote Marine Lab, about two thirds of the way down the Florida Keys. This facility is now a state-of-the-art coral growth facility, featuring seventy-four indoor and outdoor "raceways," long tanks of flowing seawater with perfect conditions for coral growth, as well as two large underwater coral nurseries.

Around the same time that Dave had his revelatory conversation about coral, Ken Nedimyer, a Florida fisherman and diver who also dabbled in the aquarium trade, had his own eye-opener. One of his businesses was growing live rocks, which are rocks that contain beneficial bacteria and other microorganisms that support the health of aquaria. They are made by simply leaving rocks out in the ocean to accumulate a natural contingent of bacteria. In his live-rock farm, Ken noticed six staghorn corals growing on his rocks. That was odd because all around Florida, the corals had been dying. So, rather than move or break them off, Ken decided to see if he could nurture them. When Ken's daughter Kelly needed a 4-H project, and after talking to some federal biologists about developing a nursery, they decided to see if they could turn those six staghorn corals into many more. The underwater orchards Ken and Kelly developed are made of PVC pipe trees moored at the seafloor with concrete and held afloat by air-filled gallon jugs or other floats. Fragments of branching coral are tied to small bits of wire and hung like Christmas tree ornaments from the branches until they grow big enough either to be glued back on the reef or to be broken in two and returned to the nursery as brood stock. Ken and Kelly planted their first corals on a part of the reef damaged by a ship grounding. The original six staghorns became two thousand in just three years, and Ken continued to fragment them and replant them to recolonize the ship-damaged area. In 2007, Ken founded the Coral Restoration Foundation, a nonprofit that enlisted volunteers to grow bigger orchards, with more fragments of coral. Today, CRF is among the largest coral reef restoration organizations in the world.

CRF has become proficient at growing staghorn and elkhorn corals in their orchards, which is great for repopulating a reef with weedy thickets. However, there are drawbacks to being too focused on just a couple of species of coral. Lessons from land are informative. In the pine tree plantations across the southern United States, fires and parasites are rampant. Studies show that the bird and frog species living in monoculture forests decline by half or more compared to the original species in diverse forests. Scientists thinking about coral restorations in the Florida reefs feared that monocultures of *Acropora* corals would create similar problems for fish and invertebrates.

Besides the two dominant *Acropora* corals, the Keys are home to forty-two or so other coral species, many of which tolerate higher temperatures. They have forms known as pillar and boulder coral for their more rounded shapes. They grow slowly, majestically, plodding along at increases of a square centimeter or two each year. That ponderous growth makes them difficult to garden and hard to use for reef restoration. One day, Dave Vaughan, who was working at Mote Marine Lab at the time, was moving a coaster-sized piece of boulder coral, but it had begun to adhere to the wall of the tank. As he pulled on it, it accidentally cracked into pieces. One bit was roughly the diameter of a pencil eraser, just a few polyps. Dave expected the small scrap would soon die. But, like Alexander Fleming with his moldy petri dishes, rather than clean it up, he left it alone. The move was fortuitous. Dave was shocked to find that not only had the tiny scrap of coral survived, not only had it healed, but it had grown "like the dickens." In just two weeks, it was the size of a dime. Dave calls that moment his "Eureka mistake." Follow-up experiments show that these "microfragments," as they've come to be known, grow ten times faster than massive coral normally does.

It wasn't that fragmenting coral wasn't known before that Eureka mistake—coral growers in the aquarium trade had been doing it for years. That's how they make a living, in fact. They usually keep a "mother" coral and then break off small pieces to sell while maintaining the "mother" colony as a brood stock. Coral hobby magazines include articles on "fragging," as it's known, nearly every month. But what was different about Dave's realization was that fragging could be used, not to populate aquaria, but to repopulate the reef. And then, as if from a kind of coral Manhattan project, came the discovery of fusion. Starting with one piece of coral the size of a coaster, Dave cut a couple dozen microfragments. When he glued these stamp-sized corals on a hard surface—say, the top of an already dead coral—a few inches apart from each other, the coral bits quickly grew toward each other until they met.

Normally, individual corals in the wild compete for space at their margins. This results in mini reef wars in which corals deploy a number of offensive and defensive tactics. In addition to incredible stinging-cell-laden tentacles, some corals have special extra-long, extra-stockpiled "sweeper tentacles" that assertively patrol for trespassers. Aggressive coral might toss enzyme-laden strands out of its stomach at its neighbor to digest it externally. Coral might simply build their skeletons atop a neighboring coral, smothering an adjacent colony. Or a coral might paralyze and eat its neighbor first, then build on top of it. Territorial skirmishes on a reef are violent and constant.

But when microfragments from the same original piece of coral grow toward each other, rather than set off alarms and launch codes, the fragments recognize that they are soldiers fighting under the same flag. Despite having no brain or central nervous system, they intuit that they are confronting not a foe but a fragment of themselves. The fragments merge into a single unit, in a process called fusing or reskinning. Because of the accelerated growth of microfragments, reskinning can produce a colony the size of a hundred-year-old coral in just a couple of years.

We usually think of sexual maturity as something that comes with age. For coral, that age occurs when it reaches the size of a dinner plate. Under typical conditions, attaining such girth takes anywhere from ten years to a quarter century. However, fused corals seem oblivious to the fact that they are only two years old when they reach the size of maturity. Once they grow together and attain a big enough size, no matter how quickly that happens, they begin to spawn. "We can close the life cycle of a coral in just two years," Dave said, which is aquaculture-speak for the amount of time it takes to start with an egg and grow an adult that produces the next generation.

In the years since Dave Vaughan's Eureka mistake, an explosion of research on microfragmenting and fusion has occurred. Coral labs have been acquiring rock saws to hack coral to bits, bits that are able to grow at speeds no one ever predicted were possible. Mote Marine Lab has refined the microfragmenting process, and their numbers show what a difference it has made. Between 2009 and 2011, the lab had been able to produce 130 boulder corals a year. But between 2013 and 2015, they produced 3,646 microfragments per year. That's a gain of more than 400 percent per centimeter of coral. Experiments showed that after a few months of growth in the lab, microfragments survived on the reef just as well as larger pieces of coral. At the start of 2019, Mote reported growing 32,000 fragments of four different species of coral in its raceways. That same year, 4,505 corals that got their start as microfragments were returned to the reefs. In 2020, observers watched as replanted coral began to spawn, evidence of successful readjustment to the wild. As for the Coral Restoration Foundation, today their orchards nurture more than three hundred genetically different corals of eleven different species, making it the largest underwater ark for coral in the world. Both the Coral Restoration Foundation and Mote Marine Lab reached the milestone of planting more than 100,000 fragments back on the reefs in Florida.

14

Outbreak

When it opened in 1914, the Panama Canal, connecting the Pacific Ocean to the Atlantic, was large enough only for tankers carrying a maximum of five thousand containers. But in 2006, Panama held a referendum and increased the size of the locks in the canal. The improvements more than doubled the maximum size of the ships that could make the passage to those carrying twelve thousand containers and increased the maximum width of ships from 106 to 161 feet. This spawned a slew of retrofits in ports up and down the east coast of the United States to increase capacity for the new, larger vessels, named, as if for a sequel to a futuristic thriller, Neo-Panamax.

In Miami, the $205 million expansion to deepen 2.5 miles of the port's ship channel and accommodate the Neo-Panamax vessels began in 2013. The shipping lane was slated to cut through a staghorn coral reef. Since the species was designated as threatened under the Endangered Species Act, the Army Corps of Engineers was required to assess how bad the damage would be. They predicted that 3.3 acres would sustain direct impacts from the dredging. Keeping a close eye on the project was Rachel Silverstein, executive director of Miami Waterkeeper. The Waterkeeper Alliance is a nonprofit founded in New York in response to the environmental degradation of the Hudson River. It sued over a hundred polluters in the 1990s, forcing them to spend hundreds of millions of dollars in remediation. Today the alliance is spread throughout the world, keeping watch over more than three hundred waterways. On behalf of the Waterkeepers, Rachel pressured the Army Corps of Engineers to commit $400,000 to rescue endangered corals in the path of the construction before it began. The corals were collected and housed for two years at the University of Miami and then returned to the sea once the dredging was complete in 2016. But, Rachel told me when I spoke to her by phone, the effort was nowhere near enough. A peer-reviewed study looking at data from sediment traps, surveys of buried corals, and satellite imagery released in May 2019 showed that the sediment plume kicked up by the dredging extended fifteen miles from the dredge site. More than half a million corals near the Port of Miami were destroyed by the expansion.

The intersection of the location and timing of the deep dredging project (between 2013 and 2016) and the onset of the coral tissue-wasting disease I'd heard about at the Reef Futures meeting (in 2014) is hard to ignore. During the same period there was a massive seagrass die-off and a dolphin mortality event near Miami. In 2014, Miami's corals were already weakened from a bleaching event and an increase in nutrients from runoff. While there was rapid acidification from the sediments that were suspended during the dig, as well as heavy metals from overturned sediment, it's impossible to draw a straight line between the project and SCTLD, because the pathogen still hasn't been identified.

When I asked Rachel about the connection, she said, "We don't know for sure, but it's highly suspicious. The disease started in the middle of the dredging and adjacent to the shipping channel. The combination of stressors creates a weakened area where a disease could take hold, like wildfire spreading through dry kindling."

DESPITE OUR HIGH HOPES, the decrease in Isy's symptoms from the round of antibiotics hadn't held. Within a few days, she stopped reading again. Her handwashing shifted into longer and longer showers. She even refused to turn the television on or off, which meant Keith or I had to do it for her if she wanted to watch. And that meant she mostly didn't. She stopped using any social media, which would have normally delighted me. But this was not normal, and only further cut Isy off from the world around her.

Isy's contamination OCD had metastasized, spreading to the public high school she'd started at the beginning of ninth grade. We could barely get her to the bus stop in the morning. After she arrived at school, she would repeatedly call Keith and me in an agitated panic. Almost daily, the school counselors would find her sobbing in the hallways. She was hardly attending class, much less learning anything. After a miserable semester, we switched her to a new school, a small, alternative learning environment where the classes were more like discussions and the teachers more like educational guides. Because everything was face-to-face, she could dodge the problem with lying about doing homework. The one class that required tests was an online geometry course, and she'd discovered a work-around for her fear of plagiarism. If she took repeated screenshots of her work before and after each completed question, she could prove to herself—or to her OCD—that she was doing the work herself. The sound of her schoolwork was the repeated grind of the simulated camera click, a noise that came to feel like a blade etching away at my heart.

One morning, I dropped Isy at her new school and went directly to a nearby restaurant to tune in to a webinar about the spread of the coral disease in Florida. I ordered a cup of tea as I logged on and fixed my headphones in my ears. Andrew Bruckner from the Florida Keys National Marine Sanctuary was introduced first. His disheveled blond hair made him look like he'd recently jumped out of the surf, though the wire-framed glasses perched on his nose added a scholarly edge. Andy has had a long career in coral protection and was an instrumental part of the effort to place the first two species of coral—elkhorn and staghorn—on the endangered species list. He had been tapped as the lead for the research and epidemiology team fighting the disease wreaking havoc on Florida's coral.

Andy began by defining the plague, which had officially come to be known as stony coral tissue loss disease, or SCTLD—pronounced skittle-dee—a discordantly carefree acronym that only heightened the grim update. As I'd learned in Florida, SCTLD was highly infectious and extraordinarily fatal. There had been some hope that small corals would escape the epidemic, Andy said, but that hadn't panned out. The disease progressed through the community in the same pattern when it reached a new area, first claiming the maze coral, then the elliptical and star corals, and finally the brain coral. Regardless of species, the animals tended to die in a few days to months. The disease ate away up to two inches of coral tissue per day.

Andy flashed an image of a diseased coral on the screen. The lesions always started near the coral skeleton and progressed outward. I could see where it looked as if the cells had dissolved, leaving only a smudge. The cause of SCTLD remained a mystery, Andy said, and it still does as I write this. Three things were known—though the clues were as helpful as knowing a gunshot victim had been killed by a bullet: the pathogen was a living organism; it could be spread either by water currents or by direct contact, and in some cases antibiotics could be used to stop it. Unusual crystals discovered in the lesions might provide another clue, but the leads were thin. Four hundred different species of bacteria were under scrutiny. Narrowing down the possible perpetrators was a slow and painstaking job. And time wasn't something anyone could afford to waste.

Dana Wusinich-Mendez from the National Oceanic and Atmospheric Administration's Coral Reef Conservation Program was also on the call. Though she had a warm smile, her voice sounded tired, like an emergency room doctor who had been up all night. Dana flashed a map of the Caribbean on the screen, and I felt myself gasp. A woman rolling silverware inside cloth napkins in the restaurant looked over at me with concern. I waved her off with a smile, though my heart sank.

The disease, which just months earlier at the Reef Futures meeting had been marching its way down the Florida Reef Tract from Miami and into the Keys, had broken free from the United States, infecting the borderless seas beyond. Red markers showed confirmed outbreaks. The northern coast of Jamaica and the east coast of Mexico's Yucatán from Cancún to the Belize border were marked. On the northern coast of St. Maarten, the disease had infected or killed 70 percent of the corals near a marine protected area. In St. Thomas, the disease was confirmed along much of the southwestern region. SCTLD had reached the Dominican Republic, where it began by attacking four species of maze coral, following its deathly pattern of infection.

The final speaker was Maurizio Martinelli from Florida Sea Grant. He was tasked with the job of describing the response to the outbreak. Wearing sleek European-framed black glasses, Maurizio described himself in one press release as "a New York city slicker who fell in love with the ocean." Rather than science, his background was in international policy, and he had worked with the UN High Commissioner for Refugees, an unfortunately apt experience. It wasn't long before I saw a headline describing the "reefugees" of the Caribbean, the healthy corals rescued and placed in aquaria ahead of the epidemic's march. Maurizio praised the many groups—at least sixty of them—that had joined together to fight the disease. He laid out an organization chart jammed with boxes and lines. Ten teams had been assembled, Maurizio said, to address the problem from all angles. He went on to describe the responsibilities spanning reconnaissance to rescue; from research and epidemiology to data management; from regulations to communications and citizen engagement. The effort was organized but urgent, defined and tactical.

Maurizio wrapped up with a slide showing the funding to fight SCTLD. For 2019–2020, Florida had mobilized just over $2 million, an amount roughly matched by the federal government. As the third-largest reef in the world, the Florida Reef Tract is 360 miles long and roughly 4 miles wide. I don't know how many coral colonies are left in Florida, but I did a quick calculation. If there's a colony every square meter and a half that's vulnerable to SCTLD, that means the state and federal government spent two tenths of a penny per coral. Earlier in the same week, the U.S. Geological Survey had released a study valuing the Florida reef at $675 million in flood protection in a regular year. In a year with a Category 4 or 5 storm, that number leapt to $1.6 billion in Miami/Fort Lauderdale alone. According to NOAA, economic activities associated with the reef in Martin, Palm Beach, and Miami-Dade Counties support thirty-six thousand jobs and generate $3.4 billion per year. The Florida reef's value is between $1.3 billion and $2.3 billion per year. At $4 million, the response to an unknown disease stripping the reef of its living tissue and marauding beyond our national borders into the Caribbean was little more than a rounding error.

At the end of his presentation, Maurizio flashed a note at the bottom of the screen in bold red font: "Not included: Staff time!" The people fighting this battle wouldn't be paid an extra cent.

ALMOST EVERY EVENING after school for six years, my son, Ben, could be found in a narrow racing shell gliding on the smooth waters of Lady Bird Lake, which runs through the center of Austin. At the end of his senior year, his crew team won regionals and qualified for a national competition in Sarasota. Keith and I flew to Florida to see the races and cheer him on. After a humid morning watching the elegant synchronicity of oars splashing, I arranged to meet with the coral aquarist at Mote Marine Lab's Sarasota Aquarium. We squeezed through a doorway into the inner workings of the facility, a place that looked like the backstage area of a theater. We skirted wooden boxes and rounded narrow corners until we reached the tank I'd asked to see.

It was three feet high and perhaps two feet deep, maybe four or five feet across. It was opaque in back and mended in places, not artfully but functionally. Two small, blue lights cast their indigo rays into the water. I squatted down to look through the front of the tank, which was dotted with snails, their greenish feet surrounding a rasping tongue that licked the glass clean. Resting in the bottom were perhaps two dozen corals, each half a foot or so in diameter. They were the sober colors of Caribbean coral: olive greens and muted purples, brackish reds and muddy yellows. These were a few of the animals that had been rescued ahead of the SCTLD epidemic, cut from the reef with chisels and hammers, sealed in foam coolers, and driven across the peninsula of Florida to rest here, in this several hundred gallons of clean seawater. They had left behind their parents and grandparents, their siblings and cousins with whom they had shared the seas. Out there, I thought, their relatives would die, and quickly. But these corals had been chosen by us. They were fated to be the survivors. These corals were simply in the right spot at the right time, near a mooring perhaps, growing in such a manner that a chisel could easily find purchase. Like a pair of animals marched onto Noah's ark, or passengers in a spaceship carrying the last of the human species off an imploding planet, these few corals might be the future of their species.

15

Lesion by Legion

One sunny spring weekend, I drove three hours north from Austin to Dallas for an event called EarthX. The expo had started as a humble Earth Day celebration, but around 2000 it caught the eye of an unlikely champion: Trammell Crow. The name itself is legend around Texas, where I'm used to seeing it plastered on the sides of buildings. Crow's father, who had the same name, had built one of the world's largest real estate companies and in the process changed the skylines of entire cities including Dallas, San Diego, and Charlotte. The younger Trammell Crow went to work in the family business, but credits time at the family ranch with fostering in him a deep love of nature. Around 2006, he founded Texas Business for Clean Air, which opposed the construction of eleven coal-fired power plants in East Texas. Then Crow revitalized the city of Dallas's Earth Day celebration. Coming from a conservative background in a deeply conservative state, Crow's philosophy is to bring everyone into the environmental tent. In so doing, Crow has become an unexpected broker between the right and the left, and the Earth Day exhibitors reflected a diverse mix of industry and environmental groups.

At the expo, I wandered from the Mexican pavilion, where I chatted with a woman about traditional embroidery, to an exhibit on birds of prey where I'd stared into the deep black eyes of a peregrine falcon. I moved on to an interactive display about beavers, where water gurgled through a mini stream slowed by mini dams, and petted a pelt so soft and dense it seemed to pull my hand along on its own. At another booth, I was handed a sample of a coral-reef-safe sunscreen. Nearby, I stopped to speak with the park ranger at the display for the Flower Garden Banks National Marine Sanctuary. About a hundred miles offshore from Galveston, salt that was left by the evaporation of an ancient sea has been pushed upward through cracks in overlying heavy sediments, creating what are called salt domes. The tops of some salt domes reach within about sixty feet from the sea surface, shallow enough for sunlight to reach coral's symbiotic algae. And there, massive reefs grow like redwood forests. Hammerhead sharks and manta rays make stops at these undersea oases. New legislation had been submitted to increase the number of domes protected from three to fourteen, and in January 2021 it quietly passed. The National Oceanic and Atmospheric Administration expanded the protected area from 56 to 160 square miles. When I asked about the condition of the corals, the sanctuary's outreach coordinator told me that, because they were so isolated, the Flower Garden Banks had largely escaped the effects of pollution and disease that plagued other reefs. Overfishing and anchor damage were controlled by sanctuary protections. However, warming waters were inescapable. "There used to be a bleaching every fifteen years or so," she said. "But there've been two in the last ten."

I asked her if they'd seen any stony coral tissue loss disease in the park.

"Not yet. And we hope we never do. But those guys over there"—she nodded three tables to her left—"they've just come back from working on it. You should go talk to them."

Down the aisle, military-looking flags adorned with golden insignias stretched across the back divider. But rather than an American eagle in the center, a blue angelfish was ensconced in the words "Force Blue." As I drew closer, I could make out the motto that encircled a twisted rope: BUOYANCY • BELONGING • BETTERMENT. Behind the table, sporting crew cuts and with their shoulders back, stood three very serious men.

I introduced myself to Jim Ritterhoff. He shook my hand with the solidity you'd expect from a member of the U.S. military's special operations. Force Blue, Jim told me as he pulled out two folding chairs so we could sit down and talk, was a nonprofit organization made up entirely of people like him, veterans who had served as combat divers. They were trained to work underwater under the most stressful conditions. Force Blue retrained these divers to put their extraordinary skills to work in the wake of hurricanes, like Maria and Irma, and recently in the face of SCTLD.

Jim turned a monitor looping through a video of Force Blue activities toward us and narrated. Force Blue divers had been asked by scientists at Nova Southeastern University in Florida to treat corals infected with SCTLD. "We deployed in teams of five, just like in combat, and were underwater from nine a.m. to five p.m." The video showed squadrons treating the infected corals. They chiseled trenches, creating a barrier between the healthy and sickened regions. They filled those trenches with epoxy and either chlorine or an antibiotic to keep the disease from spreading. "It's like firefighting," Jim said. "You drop in and build a firewall."

A firewall on a centimeter scale, I thought.

Keith Sahm, who cofounded Force Blue with Jim, joined our conversation. Like Jim, he was tall and imposing, though his jawline was softened a bit by a graying beard. "But some of them have never seen a fish before doing this work," he said, gesturing to the divers in the video.

"What?" I asked, surprised.

Jim said that special ops divers are trained to dive at night, with no lights. To do a job and return. Combat divers focus on their mission; they don't stop and admire the wildlife.

The two veterans told me how Force Blue had come into existence just a handful of years earlier. In 2015, Keith and Jim had been friends from the service, and Keith was living in the Cayman Islands, where he ran a dive shop. Jim decided to visit and, just before leaving, ran into another friend, Rudy Reyes, who had served in the Marines' special operations in

Iraq and Afghanistan. Rudy wasn't doing well. He was depressed and suffered from PTSD. "Come to the Caymans," Jim told him, and Rudy agreed. On that first dive in sunny Caribbean waters, as he saw the coral, the fish, the colors, and the reef's magnificence for the first time in his life, Rudy's world was transformed. "He was like a little kid again," Jim said.

That night at the beachside bar, the three men talked about the high suicide rates among veterans, especially in special ops returning from Afghanistan and Iraq. They discussed the heartbreaking pattern they'd seen in their friends: divorce, drink, drugs, and death. In special ops, men are trained to work together as a team toward a goal. In civilian life, finding that camaraderie and drive isn't easy. Its absence is often the void that leads to the pattern of depression and substance abuse. Compared to war, an office cubicle, a commute to work, even a family dinner can feel vacuous. Because of Rudy's experience that day, they sensed that the reef could have transformative power, one that could help other struggling veterans. "We have veterans who say, 'If it weren't for this, I'd be dead,'" added Keith. "We call it a mission therapy program." And that power could be amplified if there were a benefit to the marine ecosystem as well. They could be what they were at their best: warriors, and their common enemy would be all that threatened the sea.

Force Blue designed itself to function like it would on the battlefield. Men work in teams like squadrons. Restoration projects are treated like deployments. "We bring a 'one team one fight' philosophy to all we do. And that's different from the many scientists." Keith pointed out that scientists are often required to be competitive with one another, to vie for grants and publications. "You can't bring that attitude with us. If you go off by yourself, you get a bullet in your back. If you work as a team, you can do a lot more." And they had. In just five days, the Force Blue team fighting SCTLD in Florida had treated a thousand corals and more than five thousand lesions. In 2019, before Super Bowl LIV, Force Blue teamed up with the National Football League to restore a football field–size portion of reef near Miami, which was hosting the game in 2020. "We can be a weapon of mass construction," Jim said. "We train divers to see the coral community as a community that can't save themselves, so it's our job to go in there and do it."

Noting that the disease was spreading so fast and so broadly in Florida, I asked Jim whether he thought the effort was worthwhile.

He looked at me evenly before answering. "No one is going to want to live in the Keys without coral, and then no one is going to be able to live there without coral. This is the place to bring the most attention to the problem. The only long-term solution is getting as many people on board as we can, and Florida is the place to do it."

I heard something in Jim's voice that was very different from what I was used to. So often, corals were discussed in terms of global rates of decline, in terms of losing ecosystems across the planet, and in terms of the root problem of global warming. Force Blue wasn't about the planet; they were about the country. "We can bring a different audience than science can to the problem. We can influence legislators in a way that scientists can't. Some won't meet with environmentalists, but they will meet with us. We can get both sides to understand you don't need to be a tree hugger to care about coral." To Jim, the survival of coral was patriotic.

I have long puzzled at how close the two words *conservation* and *conservative* are to each other. Just a couple different letters lead to such vastly different meanings, such massive philosophical barriers, especially when it comes to the environment. If I arrived in the United States without any knowledge of the long history that led liberals to generally support the environment and conservatives to ignore it or worse, I'd have expected the values to be held in exactly the opposite ways. Conservatives would be slow to condone environmental change or stand by as ecosystems are decimated. Liberals would support modification, come what may, assuming that human ingenuity will solve whatever problems arise. I wondered, as I spoke with Jim and Keith, just how the budget for saving our nation's coral might change if the fight was neither about being conservative nor about conservation, but simply about our shared duty.

WHEN I ASKED ISY'S PSYCHIATRIST if it would be safe for her to learn to scuba dive, both because of her OCD and because of the antianxiety medicine she was now taking, her guidance was vague. Almost no published research has been done on diving and mental illness. No studies on the antianxiety medicine she was taking and scuba diving in adolescents were available. She cautioned me to be aware that anxiety can increase in stressful situations like diving.

Nonetheless, with all the talk of coral in our house and the possibility of visits to some of the reefs I was researching, Isy wanted to take a certification class. Because she'd basically stopped reading anything on paper or turning in physical schoolwork, I was concerned about her ability to complete the training course. But as she hadn't shown much of an interest in anything new for so long, I made a few calls. I found a program at a nearby dive shop that offered an online course for the classroom part of the training, which allowed her to perform her screen grab rituals. The video lessons she took at home were a maddening click-snap of her struggles on repeat. But when I picked her up from the in-person lessons in the dive-shop pool, she was relaxed. She exuded a calm that eluded her elsewhere. Before the final written test, which had to be taken in person with paper and pen, she sobbed among the racks of wetsuits in the dive shop. But then she summoned the strength to scratch out the answers and turn it in for a grade. All that was left for her to be scuba certified were dives in the murky lake east of Austin. The instructor, a burly man who dove commercially repairing dams and docks, pulled me aside. He said that she would be the only student testing the coming weekend. "Could you join us at the lake so she has a buddy to work with?"

I agreed, still concerned about how Isy would do underwater and knowing the refresher would be good for me too.

At the lakeshore, Isy geared up like an expert, ignoring the heat and the weight of the equipment as we waddled into the water. We submerged into low visibility, and I breathed slowly to push down my own buzzing nerves. But Isy looked serene. As the instructor led her through a battery of skills—taking off her mask and putting it back on, removing her air supply from her mouth and finding it again, even taking off her buoyancy vest and tank—she showed competence and confidence. Surrounded by water, swaddled in pressure, forced to breath intentionally, and having her visual field narrowed by her mask all seemed to create the space for her to swim away from the mental problems that plagued her up on land. These were, I realized, the exact same techniques her therapists had been trying to teach her to help with her anxiety. I recalled the men in Force Blue and the transformative power of diving for the soldiers. It seemed that Isy also found something underwater that she struggled to hold on to above.

During the last dive, the instructor told us we could bring hot dogs to feed the fish. Resting with our knees on an underwater wooden platform, I handed one to Isy. A flock of bluegill and perch mobbed her, nearly obliterating her from my sight as they tore at the treat. I worried that it would be too much, scary and overwhelming. But as the swarm of voracious fish cleared, all I saw was happiness in her eyes.

PART IV

SULAWESI

16

The Tragedy of Scale

According to one economic theory, today's problems with our seas reach back beyond politics and policy to our very human nature. In 1968, an economist at the University of California at Berkeley named Garrett Hardin gave a talk to the American Association for the Advancement of Science with the catchy title "The Tragedy of the Commons." Drawing from an essay published a century earlier, in 1866, Hardin described a grassy pasture on which herders could happily graze their cows. This mind experiment occurred before what we think of as modern times, so tribal wars and poaching, along with disease, kept the number of cattle in check. Inevitably, civilization arrived and with it came formalized trade. With access to a market, each herder asked the question: Should I add an extra cow to my herd?

On the pro side, the extra cow represents extra income. On the con side, that extra cow will graze some of the shared pasture. But, each herder thought, the con is not a negative for me alone; it's shared among all herders. So the con is a just fraction of the value of the pro. On balance, the pro of getting a cow is bigger than the con. The problem was that all the herders, being rational people, made the same calculation. The pasture became overgrazed and all the cows went hungry. Hardin concluded, "Freedom in the commons brings ruin to all." The parable of the Tragedy of the Commons is obviously not just about some herders and a pasture, but about any shared resource and humanity's apparent inability not to overexploit it. Trapping of beavers and foxes, overgrazing of the lands of the American West, overuse of national parks, overexploitation of marine fisheries, and the near extinction of whales are all examples.

So is pollution. The calculus is slightly different, but the result is essentially the same. Rather than removal of some shared resource, like grass from a commons, pollution involves adding things—chemicals, radioactivity, and, though it hadn't yet become apparent to Hardin in 1968, carbon dioxide. Everyone shares the cost of discharging wastes into the common space. It's cheaper for individuals to pollute.

When Hardin wrote about the Tragedy of the Commons, power plants were spewing sulfates and nitrates into the air, causing acid rain. Cancer-causing chemicals like DDT were being dumped into rivers and soils. As a nation, we solved these tragedies of our commons through a raft of environmental measures. Passed in the early 1970s, the Clean Air and Clean Water Acts went a long way toward maintaining the commons that are our shared air and water, a legacy we still enjoy today. For too long, we have failed to recognize CO_2 in this same way, creating a legacy that will haunt our children. The future of the coral reefs requires reckoning with the Tragedy of the Commons. The major problems for coral—carbon dioxide pollution causing warming, fertilizer and sewage runoff causing disease, coastal development causing sedimentation, illegal fishing—all result from the calculation that depleting the commons of the seas costs less than either not polluting or restricting what is taken out. Didn't we realize we could overexploit the oceans?

The answer is no. For most of human history, the seas weren't viewed in the same category as a Hardin-like commons. A long Western cultural tradition held by both the scientific and the fishing communities saw the seas as so rich and boundless they could never be depleted. The seas were too big to fail. But, of course, they did. In the 1930s and 1940s, California's sardines were the largest fishery in North America, as John Steinbeck documented in the bestseller *Cannery Row*. But in the 1950s, the sardine catch numbers started dropping. By the 1970s, sardines were so scarce that a moratorium was placed on the fishery. In 1992, the cod fishery off North America's East Coast, which shaped both the culture and the economy for five hundred years and was long considered invincible, collapsed. Hauls fell from historic catches of 100,000 metric tons to 50. That same year, red snapper on the Gulf of Mexico plummeted. In the same decade, orange roughy tanked. In the next decade, Pacific salmon collapsed. In some of these fisheries, environmental stresses like El Niño played a role. In every case, however, the ultimate culprit was overexploitation.

Even amid the collapse, western fisheries scientists focusing eastward toward the Atlantic still supported the paradigm that the sea's resources were in a category all their own, not subject to the Tragedy of the Commons. But Hawaiian fisheries ecologist H. Scott Gordon, looking at the Pacific, proposed that the seas might have limits, writing that "the plight of fishermen and the inefficiency of fisheries production stems from the common-property nature of the resources of the sea." He also noted that on some Pacific islands, traditional systems for managing resources might be an antidote.

One bright sunny winter morning in Austin, I arrived at the home of Lizzie Mcleod. We'd planned to meet at a nearby restaurant, but she needed to be with her daughter, and rather than cancel, Lizzie asked if we could meet at her house. I agreed and picked up a bag of breakfast tacos on the way. A mutual friend described sitting down with Lizzie like catching a falling star, and with a falling star within my grasp, I didn't want to miss my chance. Tall with dark, curly hair and a sparkling smile, Lizzie is the global reef systems lead at The Nature Conservancy. This, the largest environmental nonprofit in the Americas, plays one of the biggest roles in supporting the health of coral reefs globally.

One of the most critical things she'd come to understand in her years studying the reefs, Lizzie said as we peeled back the foil on our tacos, was that she needed to learn from the people who lived near them. She told me about the practice of *sasi* in Raja Ampat, a group of islands in Indonesia smack dab in the middle of the Coral Triangle. The seas' great center of biodiversity, the Coral Triangle is home to nearly six hundred species of hard coral, which is between two thirds and three quarters of the total, and at least two thousand species of reef fish. There, the rules of *sasi* had been enforced by traditional leaders for centuries. In one village that Lizzie had studied, elders gathered in a ceremonial circle around a stone and a shell and took vows to abide by the rules of *sasi*. To initiate its start, they built a "*sasi* tree" from pieces of bamboo tied together in the shape of an X, and decorated it with flowers, betel nuts, and leaves. They placed the tree on the beach, an indication that *sasi* was in effect. During that time, it was forbidden to cross the tree to take from the sea. A council of local leaders administered punishments for violations that often included fines used to improve the village. When Lizzie asked about the origins of the practice, villagers said they believed that *sasi* had been developed by their ancestors to protect marine resources for current and future generations.

In almost every region of the Pacific, traditional owners had developed such systems for protecting local marine resources, which as a group are called "traditional marine tenure." Details varied from place to place, but could include which species could be caught, when they could be harvested, and how many could be taken. Regulations might control the gear that could be used or the size of the fish that could be caught. Some places imposed limits on taking bird eggs, because local fishermen followed seafaring birds to find schooling fish beneath the sea surface. In other places, chiefs banned taking fish and shellfish from lagoons, keeping those more accessible stocks for times when the weather was too rough to head out to sea. Restrictions on taking turtle eggs, taking turtles from the beaches where they were most vulnerable, and from visiting nesting sites were common. As tropical ecologist Robert Johannes, one of the first to describe traditional marine tenure in 1978, wrote, "Almost every basic fisheries conservation measure devised in the West was in use in the Pacific centuries ago."

It might have taken the West longer to get there, but today we do have systems of regulations for our seas. In the last couple of decades, a concerted push to create marine protected areas, or MPAs, has spread throughout coastal nations. Like national parks on land, MPAs protect the marine life living within their borders to various degrees. The rules are a patchwork, some simply regulating the types of vessels that can enter the area, or their speeds, others preventing the take of some species but not others; the most restrictive and fewest in number—called "no-take" MPAs—prohibit all extraction. Significant effort has been placed on increasing the area of the oceans enclosed by MPAs. Before about 2007, less than 0.1 percent of the global ocean was protected by MPAs. By 2021 that number grew to 7.7 percent, 2.7 percent being no-take. According to the National Oceanic and Atmospheric Administration, there are more than seventeen hundred MPAs in the United States and its territorial waters. They cover 41 percent of the U.S. marine waters, although the vast majority of that high percentage is one MPA. Papahānaumokuākea Marine National Monument, which includes the Hawaiian and Midway Islands, is twice the size of Texas. But as more MPAs are designated, there's concern that MPAs might not be the silver bullet we'd hoped. Sometimes MPAs offer only marginal protection. Funds for enforcement are often lacking, and when local communities who live near MPAs aren't involved in their management, MPAs often fail. MPAs may be designated in places that don't protect critical habitat. These cases have come to be called "paper MPAs" because although they are designated on paper, they show little benefit.

Lizzie told me that even with their problems, MPAs have continued to gain prominence as a conservation tool. In 2019, the United Kingdom started an initiative to increase global MPAs from an expected 10 percent of the sea in 2020 to 30 percent by 2030. The effort gained traction, and the United Nations incorporated the idea into its goals in 2020, officially calling it 30x30. But getting there won't be easy. To achieve 30 percent protection, some countries will need to carve MPAs into their exclusive economic zones. Also, some of the rich fishing grounds around Antarctica will need to be protected. Russia and China have so far blocked those efforts. Lizzie said that these topics would be the subject of some of the biggest international environmental meetings in the coming years. And that was just about the moment we both noticed the air around us filling with smoke.

Lizzie dashed in the kitchen to see the microwave sputtering. Her daughter had gotten hungry. Hoping not to bother us, she'd put a taco in the microwave still wrapped in tinfoil. Not missing a beat, Lizzie hit the stop button, pulled out the smoking taco, plopped it in the sink, and said, "Also, there's no protecting any region on a map from the accumulation of carbon dioxide in the atmosphere and the warming of the seas from climate change."

AFTER MY CONVERSATION WITH LIZZIE, I thought a lot about the implications of traditional marine tenure. If there were places that managed to skirt the Tragedy of the Commons through systems like *sasi* in Indonesia and no-take MPAs, what does that mean for the validity of Garrett Hardin's idea? Is the Tragedy of the Commons a fundamental truth of humankind, or is it flawed?

Flawed, was the answer articulated in the 1980s by Elinor Ostrom, the first—and until 2019 only—woman to have won the Nobel Prize in Economic Sciences. Known as Lin, Ostrom was rejected from the PhD program in economics at UCLA because she'd never taken trigonometry. That would turn out to be their loss. She reapplied to the political science department and began working with Vincent Ostrom, whom she eventually married. He was studying the politics of the L.A. aquifer system and asked Lin to research the system called West Basin. The competing demands of farmers, builders, and residents were draining it dry. Observing the water wars and pumping races gave Lin critical insight into the Tragedy of the Commons. Remember the herders grazing in Garrett Hardin's field? Lin recognized that for Hardin's logic to work, he had to impose unnatural rules on the herders. The herders' pro vs. con calculus depended on their never speaking to one another or having any preconceived notions of what the other herders might do.

That's not realistic, Lin argued. "Why should we expect perfectly rational individuals placed in highly irrational structures, with no opportunity to change the structure, to achieve collective rationality?" She pointed out that it was irrational to assume "no communication among participants, no previous ties among them, no anticipation of future interactions, and no capacity to promise, threaten, cajole, or retaliate." Speaking to one another about our problems is basic to human interactions.

Lin Ostrom scoured the world for examples of humans managing commons. She found failures, tragedies of the sort Hardin articulated. But she also found successes, which she studied carefully. In Turkey, she studied local fishermen who shared their fishing grounds in a very prescribed manner. In Japan, locals in several mountain villages developed rules for sharing the forest resources. The Pacific's traditional marine tenure systems also fit the pattern.

These exceptions were fodder for Ostrom's rewrite of Hardin, which she and her colleagues called the "drama of the commons." Successful commons share one similarity: the people involved communicated about shared resources. They designed their own rules for managing their commons and generated trust and respect among one another. Recent research bears this theory out. A 2018 study of twenty-seven MPAs from around the world found that "stakeholder engagement was consistently selected as the most important factor affecting MPA success; its absence was most often linked to failure."

Lin Ostrom discovered a critical flaw in Hardin's tragedy, but her work had a problem as well. Despite all her looking, she never found a successfully shared commons the size of a nation, or even a state, not to mention an ecosystem as big as the coral reefs. And this is the Achilles' heel of Ostrom's drama the commons. Buy-in had to be created with face-to-face communication, which is hard to scale. But Ostrom left us with insight into exactly what that challenge to overcoming scale is: convincing people that they have a stake in the game.

And for that we might need to turn to the people who understand human behavior even better than economists: businesspeople.

17

The Scale of Tragedy

At the Reef Futures meeting in Florida, I heard about one of the biggest and most successful coral restoration projects in the world—according to the presentations I'd seen, anyway—which had been in Sulawesi, Indonesia. However, there wasn't a public institution or a nonprofit foundation behind it. The visionary behind that restoration project was Frank Mars, the great-grandson and namesake of the founder of Mars, Incorporated, which is much more than a vendor of sweets. It is also a major pet care company that includes Pedigree, Whiskas, and Sheba, among others. Founded in 1911, Mars is still owned by the grandchildren and great-grandchildren of the original Frank Mars. According to Forbes, it is the sixth-largest privately held company in the United States, employing over one hundred thousand people. In 2019, its revenue was $37 billion, and the Mars family ranked as the third-wealthiest family in the country, after the Walton and Koch families. That most of us aren't aware of the Mars family, that their names don't plaster the headlines, is intentional. A 1992 article in the UK's *Independent* is one of the few profiles of the family. It describes Frank's father, John, driving to Mars headquarters in Virginia in a three-year-old Jeep station wagon and clocking in on a time card just like any other employee. There were no executive suites, no reserved parking spaces, no oil paintings of the dynasty hanging on the walls. The Mars family even declined to be photographed for the story. The modest, low-profile leadership style seems to serve a purpose: Mars is consistently ranked as one of the best places to work in the United States.

After the Reef Futures meeting, I reached out to Jos from Mars, the company's sustainability manager. One day in late spring, he wrote to tell me that Frank Mars would be visiting Austin and had offered to meet me in person to talk about coral restoration.

At the front door of a classic dark-wooded steakhouse in downtown Austin, Frank and I shook hands. He was a large man with small round glasses and thinning gray hair. He exuded the air of a serious business-man, but once we were seated I found him easy to speak with. After we ordered and talked a bit about what brought him to Austin (a visit to one of the biggest Skittles, Snickers, and Starburst factories, in nearby Waco), I asked my most burning question: Why has the Mars company developed a coral restoration program?

Frank told me the story: In Sulawesi, Mars had a factory that pro-cessed cocoa into powder and other products. As in many places adjacent to coral reefs, fishing is one of the most important sources of protein for Sulawesi's residents, and local suppliers had struggled with the fish get-ting smaller. At the same time, the Indonesian government had initiated an effort to develop alternative sources of income for the people who lived near the reefs. Collecting seahorses for the aquarium trade had been one of those income-generating ideas. But even the seahorses were disap-pearing. Frank had been fascinated by coral reefs since he dove on one as a child. The interest grew into a passion for conservation as an adult. He knew local reefs had degraded because of poor sanitation and illegal fish-ing. He recognized that the underlying problem for both the shrinking fish catch and seahorse population was degraded reefs. He asked Noel Janetski, who ran the cocoa factory and was also an avid diver, to survey the restoration technologies available with an eye toward restoration tech-niques that were easy to build anywhere, scalable, and exportable. Fifteen years' worth of R&D later, they figured out how to do it. The system is based on a six-legged rebar structure they called a "spider." Each spider is about a yard in diameter, small enough for an individual diver to hold and

manipulate. The spiders are coated in sand to encourage coral attach-
ment, and each spider is loaded with fifteen coral fragments, zip-tied in
place. Then the spiders are networked together into an undersea web.
Frank said that a team of eight divers could build out a basketball court–
size area in three hours. "That will put up to forty-five hundred corals on
the reef." The number was impressive. For comparison, it was about what
Mote Marine Lab in Florida replanted in a year. What's more, the cost
came down to about a dollar or two per coral. Again, it was impressive.
Estimates from the Caribbean were more like fifty to one hundred dollars
per coral.

When I asked about the cost of the entire project, Frank said that Mars
spent about $250,000 per year for research and development as well as the
actual restoration. The funds came directly out of the Mars budget. Frank
then referred to the five principles on which the Mars company is founded.
"One is responsibility. I don't think we should profit from restoring the
coral reef." For the record, Mars's other four principles are quality, mutu-
ality, efficiency, and freedom. The principles don't just take up space on
the company's website. Everyone who I met who worked for Mars knew
about the principles. They came up frequently in conversation.

But, Frank said, "philanthropy alone won't solve the problem of the
corals. Nongovernmental organizations won't solve the problem. We
have to say, 'Industry, if you want your reef, you are responsible.'"

Frank envisioned creating a market ecosystem where companies chal-
lenged one another to restore reefs using the spider system. This would
bring the problem of coral reefs more attention and funding. "Every time
there's a build, you get the halo effect through publicity, and that engages
consumers," Frank said. "Employees and millennials are demanding that
companies act in a way that helps the planet."

A FEW MONTHS AFTER Frank and I met for dinner, I received an invita-
tion to visit the Mars restoration site in Sulawesi. I'd be joined by Richard
Vevers, the ad man and star of *Chasing Coral*, and his wife, Stephanie Tate,

who is a marine scientist also working at The Ocean Agency. We would have the opportunity to scuba dive on the restoration and even partici- pate in a small build ourselves. I accepted at once. And then, because the trip would fall right in the middle of summer vacation, Keith and I de- cided to splurge on a once-in-a-lifetime family trip afterward in nearby Bali, where we could dive together as a family on a coral reef.

I boarded the long, thirteen-hour leg of my flight, Dallas to Seoul, and walked down the aisle toward the window seat I'd been sure to reserve online. Never a good flier, I knew that if there was any chance of sleep during the flight, I needed to be able to pull a hoodie down over my eyes and curl up against the window. But arriving at my seat, I found, as the three bears had, there was already someone in it. And she'd made herself quite comfortable too. A small, formidable, gray-haired Goldilocks, shoes slipped off, blanket spread out with a seat belt buckled over it, and a mas- sive purse tucked under the seat. I look back and forth at my seat number and, yes, she was definitely in my seat.

"I think that's my seat," I said.

She smiled widely at me, making it clear that she didn't understand, didn't speak English, and anyway wasn't moving. She pointed emphati- cally at the aisle seat.

"No, that's not my seat. See." I held out my boarding card with the seat number on it.

She nodded and pointed again at the aisle seat.

I shook my head, pointing at the little cartoon of the seats and the window printed on the overhead compartment. My seat was clearly next to the drawing of a window, not a person walking.

She nodded and flapped her hand at the aisle seat.

I held out my boarding pass again, pointing at the seat number.

We were battling it out in passive-aggressive smiles and silent hand gestures. I needed reinforcement, so I called over the flight attendant, who confirmed, yes, that was my window seat. But this Goldilocks wasn't in the least concerned about the authority of the flight attendant's tidy uniform and neat bun. She just kept smiling and pointing at the aisle seat with great conviction. The flight attendant looked at me helplessly. The travel fates had pitted me against someone with the Garrett Hardin Tragedy of the Commons philosophy. She'd gotten to the seat first, and so she took it.

"Fine," I said, collapsing into the aisle seat. I shot the woman a stank eye that said, "Just wait until you have to pee."

She shot back a satisfied look that I was certain meant, "No problem, I'm a camel."

I plugged myself in to a modern-day Agatha Christie movie, then flipped up my hoodie and tried to curl up against the nothingness of the aisle. I was unable to settle into a comfortable position, and my mind drifted to the questions of scale. How much *would* it cost to restore the world's coral reefs?

According to the United Nations, corals cover 28 million hectares. About half the reefs are already damaged or dead, and not all reefs are candidates for restoration. So, for the purposes of an airplane napkin calculation, suppose we want to restore 10 percent of the reefs globally. That's 2.8 million hectares. A hectare is 10,000 square meters, which brings us to 28 billion square meters to restore. Frank Mars had said each of his rebar structures held fifteen colonies, and from the pictures I'd seen, they looked roughly a square meter in size. Frank also said his restoration costs were about $1 to $2 per coral. That meant restoring a tenth of the world's coral reefs with his method would cost in the range of $4 trillion to $8 trillion. Just how big is that? The EPA's entire annual budget is around a thousand times less, at between $8 billion and $9 billion. On the other hand, subsidies paid to fossil fuel industries are between $400 billion and $5 trillion per year. Or, cutting to the chase, it's estimated that it would take $2.4 trillion per year to keep climate change at bay.

Thinking about value versus expenditure, the value of coral reefs can be assessed based on the services they provide to humans, called "ecosystem services." Ecosystem services include everything from regulating the climate to storm protection; from erosion control to providing habitats for other creatures. Trying to quantify these ecosystem services bakes in a lot of assumptions, which have been duly criticized by scientists and economists. One of the typical complaints is that trying to value nature is meaningless because once it's gone, it's invaluable. But let's just suspend those objections and put some boundaries on the problem. A 2014 study of coral reefs valued their ecosystem services at $362,000 per hectare per year, twice that of tidal marshes and wetlands, which are the next closest ecosystem in value. (Surprisingly, they are also valued at seven times more than the world's other hyperdiverse ecosystem: tropical forests.) Frank had said it required a $250,000 investment to restore a hectare of reef. So you've got a return on investment of about 1.5 with coral reefs. The analysis also showed that, globally, coral reefs provide almost $10 trillion in ecosystem services to economies. In view of my back-of-the-napkin $4 trillion to $8 trillion range, coral restoration was a reasonable investment.

At Reef Futures, Richard Vevers had said he envisioned a wave of public support for coral reefs driven by advertising. Frank Mars envisioned a wave of support driven by companies whose millennial customers demanded it. As I sat in my aisle seat, trying to find a way to get comfortable, I mused that there seem to be two ways to see the world, and trust was at the root of both. According to Lin Ostrom, Richard Vevers, and Frank Mars, people think: This thing, this color, this idea you have is good, and I trust you, so I'll join you. But Garrett Hardin and the grandma in the window seat next to me saw just the opposite: This thing, this color, this idea, is good, and I can't trust you, so I'll take it before you do.

Or maybe we all have both parts in us, the trust and the mistrust, and people who want to sell us stuff exploit them both to change our behavior. They use trust to form tribes and mistrust to bind the tribes closer together. Averting the tragedy of the coral hinges on building tribes of support larger than any we've ever built for any ecosystem before. But unlike four-ply toilet paper, corals aren't an impossible product or something no one needs.

In fact, they are just the opposite.

18

A Place to Restore

The first time I looked for the island of Sulawesi on a map, I found it on the edge of the Ring of Fire in the western Pacific, where the oceanic tectonic plate dips below that of the Asian continent. It struck me as having the form of a dragon standing on its back legs, jaws open, roaring ferociously east across the equator as if helping to fuel the El Niño–heated seas. The world's eleventh-largest island by landmass, Sulawesi is a sinewy collection of volcanic ranges surrounding three enormous gulfs with nearly five thousand miles of coastline. The city of Makassar, where I was to meet up with Richard, Stephanie, and the Mars restoration team, sits behind the back knee of my imaginary dragon, along a wide shelf freckled with low-lying coral-surrounded islands known as the Spermonde Archipelago.

Perhaps it was the great tectonic upheavals of the land that drove some of the people of Sulawesi to become proficient on the seas. The Bugis people lived in southern Sulawesi as early as 2500 BCE, were skilled sailors and boat builders. They sailed east to Papua New Guinea and south along the Australian coast, trading birds, skins, and medicinal bark for pearls and dried sea cucumbers known as *trepang*. When the Portuguese and then the Dutch arrived in Makassar in the 1500s and 1600s, they called those sailors ruthless pirates. Some say they brought home stories of "Bugi men," or boogeymen. Other etymologists say the word had different origins and the similarity was just a convenient coincidence. What isn't up for debate is that the Europeans found a rich trading industry already established, which they subsequently brutally subdued and colonized. Among the many goods exported to Europe was a deep brown oil made from coconut or palm and the flowers of the ylang-ylang tree. In England, a barber named Alexander Rowland offered it to his customers as a hair tonic, claiming it might also prevent baldness. So many men poured the unctuous balm on their scalps that stained upholstery became problematic. In response, fashionable households placed a protective bit of lacework across the back of sofas and chairs. When I'd boarded my flight for Indonesia, I hadn't realized the flimsy cloth attached to my headrest was called an antimacassar, still named for the exported riches of the very city that was my destination.

When I checked into my hotel, I received a message that Noel Janetski, Mars's director of marine sustainability, had invited everyone to a welcome dinner at his house. After dropping off my bags, I headed out into the streets to buy a small gift as a thank-you. I immediately encountered problems. Technically, Makassar has sidewalks, but the unstable ground coupled with the rich volcanic soil meant that massive overgrown roots of trees crumpled the concrete every few feet. On the street side of the trees, I risked sideswiping by speeding scooters and trucks. On the opposite side, I risked falling into the open gutter. As I picked my way over the riotous ground, a barrage of toots and honks showered over me. I worried that my uncovered head and bare arms in a Muslim-majority country were to blame. At the same time, none of the shops on Google Maps were where they were supposed to be. Or maybe they were and I couldn't tell, because storefronts were masked by imposing gates. I had a bewildering sense that you had to know where you were going before you went anywhere in Makassar. I eventually bought a few cakes at a shop called Holland Bakery, a legacy of the centuries of Dutch rule.

That evening, I met Alicia McArdle, Mars's marine conservation manager, in the lobby of our hotel, and we walked over to Noel's house together. Richard Vevers joined us, offering apologies for Stephanie, who was exhausted from jet lag. Guided by Alicia, who traveled to Makassar about every six months from her home in Australia and knew her way around the challenging terrain, we chatted on the way to Noel's house. Alicia had studied marine biology and then worked for the Australian government as a conservation officer. Her previous job had been in biosecurity, including controlling the export of coral for the aquarium trade. By the time we arrived, I recognized in Alicia the capacity to be at the same time extraordinarily competent and mischievously funny.

Noel's house was set behind yet another imposing gate, which slid open to admit us. His young daughter swung open the door and threw her arms around Alicia. They'd become good friends over Alicia's many visits. Noel's daughter flashed us all a welcoming smile and said her parents would be a minute. While we waited, Alicia mixed us drinks. Noel arrived and elbowed her out of the way, mixing himself a rum and orange juice with soda water.

"Who drinks that?" Alicia said, wrinkling her nose at him in mock distain.

My first impression of Noel was one of an Australian Robin Williams in his mustache phase. He joked constantly, but the more he talked, the more his sharp intellect emerged. Noel was educated as a chemist, and by nature he was a project-oriented thinker. At some point in most conversations he would ask something along the lines of "But what's your objective?" followed by a list of several you should probably consider.

Dinner was build-your-own-pizza, and Noel ushered all of us into his dining room saying, "Pizza-making is a test to see how good you'll be at coral restoration." The long table was loaded with giant pans on which pizza dough had been stretched. All variety of toppings ran down the center. We each created our masterpieces, with a fair bit of competitiveness. I asked for the jar of tomato sauce. It was Mexican salsa instead of marinara, but, I figured, why not? I topped mine with an assortment of veggies, tomatoes, and peppers. Noel and his daughter teamed up and loaded theirs with pineapple and salami.

As we waited for the pizzas to cook, Noel mapped out an imaginary coral reef on the dinner table using salt and pepper shakers, and bottles of beer and hot sauce to represent the shoreline, living coral, and dead rubble. He demonstrated where he would install spiders for success and where he'd learned, through years of trial and error, they would likely fail. The little things mattered a lot, Noel said.

Richard asked, "Say you have access to money for reef restoration. What should you do with it?" This, I would find out later, was more than hypothetical. A client had asked Richard for advice on where to make a $10,000 donation for coral restoration.

Noel answered, "Look for places where the restoration will endure." He knew he had a real advantage working off the coast of Sulawesi. Indonesia is located at a critical juncture between the Pacific Ocean to the east and the Indian Ocean to the west. The currents that braid their way through the complicated splattering of islands, like a stream that has to find its way through a chain of boulders, is called the Indonesian Through Flow. Like water accelerating over rapids and then eddying in the back-wash, the currents of the Indonesian Through Flow are wildly variable. When Pacific water bumps up against northern Sulawesi and then passes through the deep channel between Borneo and Sulawesi, the volume of water cranks by at 11.6 million square meters a second. It then meanders around the Java Sea south of Sulawesi and exits through the Lombok Strait east of Bali at a lackadaisical 2.6 million square meters per second. The currents pick up speed as they head south, eventually flushing into the Indian Ocean at 15 million square meters per second.

What the erratic flow pattern means is that lots of coral spawn are swept into the Indonesian reefs and retained there long enough to find a new home. This both increases the quantity of raw material feeding the reef and bolsters genetic diversity. Restoration projects have a much better chance of success under those circumstances than in places where the genetic pool is shallower. During my meeting with Frank Mars, he had told me about this oceanographic benefit. He'd said, "We are working in the Indonesian Through Flow, so we get tons of seeds. There's a global reservoir there."

Noel continued with a second criterion. "You should also look for places where climate change isn't as bad." Despite the term *global* in front of *warming*, the effects aren't evenly distributed around the world. The poles, for example, are warming at two to three times the rate of temperate regions. But assessing global warming in Indonesia is complicated. The Indonesian Through Flow is embedded in a larger ocean region called the Indo-Pacific warm pool, often called the "heat engine" of the planet. How the heat engine will respond to climate change and what that means around Indonesia hasn't been well-studied, although it should be. A review of the region's oceanography concludes vaguely, "The IPWP [The Indo-Pacific Warm Pool] deserves close scrutiny and investment for a better future of humanity and the health of our planet."

But the evidence on the ground, or just below the waves, was the most compelling reason for working in a place like Sulawesi. Following the Third Global Bleaching Event in 2016, researchers surveyed 226 reefs in the Indo-Pacific to see how the coral fared. Despite some of the greatest Degree Heating Weeks in the region, rising to the top in terms of resilience was the Coral Triangle. Of those many reefs, the ones in the Makassar Strait showed among the highest resistance to bleaching. Returning to Richard's question: What should you do if you have money for restoration? Mars seemed to have chosen a place where its investment was very likely to succeed.

19

A Reef of Hope

The next morning, the call from the muezzin reverberated through my hotel room as if the loudspeaker were mounted on the wall directly above my bed. I squinted at my cell phone. It read 4:30 a.m. I pulled a pillow over my head. That first call was echoed by another, this one a bit more distant, maybe as far away as the bathroom. Then a third from somewhere around the closet. Soon, it became a chorus, the entire city of Makassar awoke in the call to morning prayer. I peeked out from under my pillow. The first bit of light glowed in the sky. The calls and echoes continued until the entire sky was lit by the sun.

A few hours later, after consuming as many small cups of the weak coffee the hotel restaurant served as I could reasonably order, Alicia, Richard, Stephanie, and I loaded our dive gear into a car and set off for the city's central port. When we arrived, the first thing I noticed were giant metal letters twice as tall as a person curved along the far end of the port's plaza, spelling out CITY OF MAKASSAR. Beyond that, a massive mosque formed from ninety-nine domes, each said to represent one of the many names of Allah, bloomed from the edge of the sea. The mosque's architecture was an intricate clustering of ever-taller cylindrical towers, colored yellow and orange like the sun. The central dome was the tallest and largest of all. I imagined that from the inside the space would be stunning.

Noel and the coral restoration team, led by Saipul Rapi, were waiting for us by a statue on the plaza. Saipul wore a floppy tan "Mars Sustainable

Solutions" hat and thin wire-framed glasses, and his dark hair and beard were both cut close. He had studied botany and marine biology in college, and besides speaking Indonesian and English, he was fluent in the local dialect, Makassarese. We boarded a small motorboat and took off to the northwest across the flat water of the Spermonde shelf. As we drove, Noel pulled out a stack of laminated maps and explained that there were about 120 islands dotting the undersea plateau. Beyond the plateau, the seafloor abruptly dropped a kilometer in depth. That, he said, waving to the west, was Wallace's line. "Do you know what Wallace's line is?"

No, I did not.

On his first sail around the world in 1521, Ferdinand Magellan brought along an Italian assistant named Antonio Pigafetta, who noticed that the animals and plants in the Philippines to the north of Sulawesi were strikingly different from those on the Maluku Islands to the southeast of Sulawesi. Several centuries later, acclaimed British naturalist Alfred Russel Wallace visited the region and confirmed Pigafetta's observations. Tigers and rhinoceros, like those found in Asia, populated islands to the north and west of Sulawesi. Kangaroos and platypuses, like those found in Australia, inhabited islands to the southeast. Wallace hypothesized that the deep, narrow channel Noel had just waved at was the reason for the differences, though, like Darwin with his hypothesis about coral atoll formation, Wallace was missing a key piece of information. The ups and downs of the ice ages weren't yet understood. When glaciers grow, sea level falls, leaving dry land exposed. Animals can walk to new habitats, and that's what happened on either side Wallace's line. Asian tigers migrated to Sumatra and Java while Australian kangaroos moved into New Guinea. But even during the ice ages, the channel just to the west of Sulawesi was so deep that it never fully emptied, remaining a barrier to land migration for fifty million years.

On the other hand, the channel has been an expressway for marine animals, Noel said. "The strait of Makassar is a tuna highway. Forty percent of all Pacific tuna brood in the Sulawesi Sea." According to a 2016 World Wildlife Fund (WWF) study called the Tuna Blueprint, the estimated value of these fish is $2 billion. However, the spawning biomass of three major tuna species has collapsed to about 20 percent of its 2007 values, largely due to overfishing.

"The question Mars is always asked," Alicia said, "is 'How much does a restored reef contribute to the tuna catch? What's the connection between the reef and a very valuable fishery like tuna?' Quantifying that number would provide strong financial justification for coral restoration." In 2018, the WWF proposed establishing an Indonesia and Philippines peace park in the Sulawesi Sea, off-limits to fishing. It would give juvenile tuna time to reach the open Pacific Ocean before they are caught, which has the potential to help their numbers rebound. But so far, that plan has remained just a blueprint.

Cruising along, I noticed similarities among the low-lying islands. Beachlines were dominated by peaked rooftop after peaked rooftop with a few bushy trees poking out here and there. The round dome of a mosque was tucked back from the seashore, and a long dock stretched out into deeper water.

Our first stop was the island of Bontasua, which was closest to the coral reef currently undergoing restoration where we would speak to officials about the project. I asked Noel why the restoration was sited there as opposed to any of the other islands.

Mars had held a sort of competition, he said, in order to see which islands were interested. A half dozen heads of islands, *kepala desa*, had applied. The first site chosen had been near a different island, Badi, about five years earlier. However, when a new *kepala desa* was elected, the relationship soured, and some of the restorations were destroyed. This was the kind of situation that Lizzie Mcleod had brought up. Without local commitment, a restoration wouldn't succeed. The restoration near Bontasua was Mars's second effort.

The boat slowed as we approached the island. Beneath, the sea was colored the brilliant aquamarine of the tropics, so clear that you could see straight down to the white glistening sand. A network of old car tires served as a ladder up to the wooden dock. As we walked down the deck, which had been beaten by the sun and elements, it wiggled beneath our feet. At the border between the dock and the island, a small wooden and corrugated-metal hut served as a shop selling chips and candy. As I glanced up at a solar-powered streetlamp, I nearly bumped into a flock of seven ducks waddling down the path in front of me.

Following Noel and Saipul, we turned right and threaded our way through narrow passageways. With no cars on the island, houses were just arm's lengths apart. Most were two-storied, made from wood or corrugated metal, painted in bright blues, greens, and yellows. At each rooftop's peak, the frame continued upward, creating a small soaring V, which Saipul said was an architectural flourish characteristic of the region. Small fenced gardens held banana palms and red-flowered bushes. Laundry hung from clotheslines, and roosters strutted along the edges of the yards.

We reached the building where official business took place and slipped out of our flip-flops, walking barefoot onto the tile floor according to the custom. Inside, three dark wooden desks were positioned along one wall. The rest of the room held blue plastic chairs facing the desks for us to use. Official posters were taped to the bold green wall along with framed photos of government leaders. One photo was an aerial view of the island. It looked like a long papaya, its densely packed houses like seeds. A breakwater trailed off the northern end and curved around to the west, like a stem. In the very center, there was a sandy area about the size of a soccer field. Beyond the island's edges, a fringing reef stretched out like a halo.

Men began arriving, mostly dressed in polo shirts and slacks, though some wore sarongs. One man's sleeve bore a logo that read STOP ILLEGAL FISHING. Four women dressed alike in dark green long-sleeved, long-skirted official uniforms stood near the back of the room. Two wore red hijabs with sparkled flowers embroidered across the top like a headband, and two wore mustard-yellow hijabs. All were beautifully made up. We smiled at one another and they each shook my hand, following our clasp with a touch to their heart and a nod of their head.

Three men took their places behind the desks. Saipul took a chair to the side of the desks and translated from the local Makassarese language. The greeting was very formal: The officials were happy we had come to the island to see the restoration. The *kepala desa* had sent word he was sorry he couldn't attend, but the others would make sure we had a good visit. They asked, through Saipul, what our goals were.

Richard first spoke about The Ocean Agency and its mission to increase awareness of coral reefs and their importance. He and Stephanie planned to document the restoration on film to let the world know that it was happening here. And then Richard asked the leaders, "How do you feel about the restoration?"

"We see it as a great achievement," said the man who was tasked with protecting the restored site from illegal fishing. "We can see that the number of fish has increased. The coral reef is both larger and more colorful. And the fishermen have more income."

Richard asked, "Are other nearby islands interested?"

Another official answered, "Yes. Most of the *kepala desa* would like to have a restoration."

After the formalities were over, I turned to look at the women in the back of the room. One of them caught my eye and smiled, her eyes framed with gold wire-rimmed glasses. A little boy in a Spider-Man shirt crawled into her lap.

WE WALKED BACK through the village and down the sun-worn dock to the boat. As we organized our dive gear, Alicia reminded us that the island had no sanitation. The bathroom was the beach. "Just beware, we've had a few code browns."

At the restoration site, Saipul cut the motor, and one by one, we flipped backward, off the edge of the boat. In the water, I released the air from my buoyancy compensator, the air-filled vest divers wear to control how much they float or sink in the water. Aside from a few practice dives with Isy when she was getting certified in the lake near Austin, it had been over a decade since I'd been scuba diving, and I was apprehensive about looking amateur in front of the group of veterans.

But when I descended, I arrived in a place that felt like home. A broad field of vibrant coral stretched before me, dense thickets of branching coral, purple and blue colonies mingling with cream- and tan-colored kin. Leafy coral, rimmed in white piping, were beginning to overshadow their cousins but provided a new habitat for fish. Schools of small blue chromides darted among the coral boughs and retreated deep into their trunks as I swam near. I saw larger fish too: long brown wrasses painted in blue stripes, narrow cornetfish, flamboyant Moorish idols, as well as so many I couldn't name. There were the little knobby knolls of the coral, their miniature polyps lending a sweet dimpling to their branches; wide-eyed large-polyp coral; golf balls of brain coral; and small fans of more slow-growing coral. And among the restored hard coral, I spotted other invertebrates: leathery soft corals waving their giant fronds, orange- and purple-stained tunicates, clumps of feathery black and orange crinoids, thin-shelled scallops, and the curvy blue lips of a young giant clam that tightened to close when I waved my hand over the top. The closer I looked, the more I saw. But as I pulled back my gaze and scanned the reef, I saw dozens of stars—healthy, vibrant reef overgrew all but the most central intersection of the rebar structures' six starlike rays. I swam by a stake and noticed a tag was attached. I scrubbed away the algae with my finger. It read BLOCK 1, JULY 2017. This reef was just two years old. Everywhere I looked, there was a vibrancy, an energy, and a feeling of growth. Most of all, there was a sense of hope.

20

A Blast in Makassar

We returned to the boat and wriggled out of our dive gear, sparkling with what we'd seen. Richard and Stephanie had taken several cameras on the dive, and Noel asked if they'd been able to capture the photos they wanted. Stephanie said she thought they had. "But I'd like to get a before shot, a place that hasn't been restored for comparison."

Saipul suggested that we stop at a rise nearby where the reef had been reduced to rubble as a result of blast fishing. "It was bombed thirty years ago," he said.

"And it's still rubble?" I asked.

"Yes, it hasn't come back. It won't come back without restoration."

Underwater, the site looked like a war zone. Everything was gray and broken, nearly devoid of life. The contrast to what we'd just seen was striking. Two researchers, Lida Pet-Soede and Mark Erdmann, who lived among the blast fishermen in Sulawesi for two years in the 1990s, published a description of what had happened: "Explosive blasts typically shatter the more delicate corals, and can leave characteristic craters in the substrate. The size and nature of these craters vary considerably with the size of the charge and distance from the substrate when exploded. . . . On a reef flat, when an average-sized bomb explodes close to the substrate, an area of 1–1.5 [meters] is often completely destroyed. Branching and foliose corals are blown into rubble, tabulate acroporids break and fall over, and larger massive corals are often cracked. If a bomb is thrown on a reef slope, the damaged area has a different shape and is usually somewhat larger due to breakage from coral fragments rolling or sliding down the slope. Reefs which are subject to repeated blasting are often reduced to rubble fields, punctuated by an occasional massive coral head."

Recovery, if it happens at all, takes decades. Baby coral, trying to gain a foothold on a piece of dead coral rubble, are crushed when the dead piece rolls around in the surge. And blast fishing isn't just a problem near Makassar. Hundreds of bombs explode every day on reefs in at least thirty countries around the world. It's possible that over half the reefs in Southeast Asia are actively bombed. Although I'd not heard it discussed before visiting Makassar, major media outlets have done embedded reporting on blast fishing in Tanzania, the Philippines, and Sri Lanka.

Historically, the people of Sulawesi practiced the kind of collective management of their fishing grounds that Lin Ostrom studied and that Lizzie Mcleod told me about. Traditional systems called *sasi laut* maintained the reef's resources for the benefit of all the fishermen. But the advent of mechanization, as well as immigration of people unfamiliar with traditional marine tenure in the twentieth century, broke down those systems. Government policies, which marginalized traditional fishermen to the benefit of large, often foreign, fishing corporations, exacerbated the need to compete on an unfair playing field. At the government level, a legacy of the bureaucracy and corruption instituted during Dutch rule remains, leading to distrust of government officials. It's generally expected that about 30 percent of the funds from every development program will line the pockets of officials. Poverty also plays a role. Even though Indonesia has the longest coastline of any country in the world, the contribution of marine resources to the gross domestic product is among the lowest, at 20 percent. Compare that to 54 percent in Japan. Per capita income in 2005 in Indonesian coastal communities was $50 to $70 per month, and the poverty rate nearly doubled between 1997 and 2009. Additionally, the educational system is weak. Seventy percent of people living along the coast don't complete elementary school. Only 0.03 percent finish high school. All of these systemic pressures leave fishermen with few choices to support themselves. Compared to hook and line or nets, which require constant repair and labor, explosives are an easier and cheaper way to catch fish.

Blast fishing was practiced as early as the 1900s in the Indo-Pacific, when a U.S. Bureau of Fisheries expedition to the Philippines used explosives to gather fish samples. Later, in the 1930s, leftover explosives from railway construction served as ammunition. But blast fishing really took off after World War II, when dynamite from the Pacific theater became widely available, often supplied by Dutch colonists. After the war, when stocks of dynamite were depleted, fishermen switched to homemade fertilizer bombs made from drinking bottles, ammonium nitrate fertilizer pellets, kerosene, and a wick. In 1998, between 10 and 40 percent of all the fish observed in the fish markets had telltale scars from explosions. In 2009, that number was up to 70 percent.

Because blast fishing is illegal, the crews become bound together as in a crime family. The captain takes on a kind of godfather role, helping crew if family members become sick or providing food if a family is in need. Crew members ask the captain for permission for their children to marry or continue with their education. Induction begins young, in the early teen years. In interviews, researcher Muhammad Chozin found that the people of the Spermonde Archipelago were well aware of the harm to coral reefs from blast fishing. Yet, he wrote, "this awareness does not seem to influence them into action."

As you can imagine, blast fishing is dangerous for the people who do it, too. Homemade bombs can explode early, and stories of people losing body parts are not uncommon. Short wicks can burn so fast that they are called *sumbu syahadat*. *Syahadat* is the prayer Muslims say before dying. Saipul told me that he knew one blast fisherman who had lost an arm but continued to fish that way, until he blew off his other arm as well. He also said that blast fishing can be particularly dangerous for women, as bomb making is often viewed as women's work.

So why does it continue? Blast fishing is cheap and efficient. In

Indonesia it costs about $1 per bomb, and the yield averages $2 of fish. Deploying three bombs a day earns a small-scale fisher what a handline fisherman makes, but with much less sweat and headache. Divers working for large-scale fishers can make as much as ten to fifteen times as much, even when paying stiff fines if they are caught. When I asked Saipul about enforcement, he said, "The police might go for them, but it's hard because the police know they are armed with bombs." Also, it's hard to catch blast fishermen because the process is so quick. "You'll see three boats pull up, and they set a bomb in the middle and then catch whatever comes up. It takes less than half an hour, and then they are gone. So if you call the police, by the time they get there, there's no one to catch."

Unfortunately, blast fishing isn't the only form of illegal fishing that poses a threat to coral reefs. Noel said, "There is also a problem with cyanide fishing." Using an air compressor or free diving, fishermen descend holding a squeeze bottle containing a potion of a few cyanide tablets mixed with water, which they squirt into nooks and crannies of the reef. The cyanide attaches to hemoglobin, blocking the spot where oxygen should attach. Without oxygen, a fish starts to suffocate, making it easy to catch alive. In 2016 more than half the fish in aquarium shops tested positive for cyanide poisoning. According to the WWF, a fisherman can earn five times more for a live fish caught with cyanide than a dead one. Unfortunately, it isn't just fish that are stunned. Any animal hit with cyanide suffocates, including coral—except coral can't move away from the poison. For each aquarium fish that's collected with cyanide, about a square meter of coral is destroyed.

OUR FINAL STOP was at Mars's first restoration site near the island of Badi. We geared up and flipped off the back of the boat into the water. Immediately apparent was that while the first site had been vibrant and hopeful, it still looked like a work in progress. Here, where the coral had been growing for between three and five years, it was glorious. I leveled my body parallel to the reef top and began exploring. In the concavity of a coral colony, I saw a school of newly hatched fish each the size of an eye-lash, and thousands of them glimmered in the sun. As my eyes scanned the reef, I saw so many fish: parrotfish, Moorish idols, wrasses, clown fish darting down into anemones. Worms with feather-duster-crowned heads in rich black and blue danced among the corals. Scallops burrowed near the bases of coral colonies. Neon-lipped giant clams kissed the sea. A long barracuda lurked on the edge of my peripheral vision. Off the edge of a crenulated cliff, I saw a big sea turtle plodding along and receding into the deep blue as if a ghost. I peered under a coral branch and saw a single Crown-of-thorns starfish, and at the sight of the predator who crippled the Great Barrier Reef, I suddenly realized I couldn't see any rebar. It was so overgrown as to have disappeared.

There are many ways to measure the health of a reef, and scientists debate them endlessly. They measure percent coral cover, roughness, species di-versity, ecosystem diversity, and more. I didn't need any of those measure-ments for this reef. I could tell this reef was healthy without pulling out a transect line or taking any photos for analysis. I knew it in my heart, which welled with a childlike joyous energy. It was the kind of sensation that can't be summoned until you feel it. It is the exhilaration of running down a hill with your arms widespread. It is the joy of watching your child do some-thing you never imagined. It is the reverence of unexpected wonder. I ca-reened down a slope of reef and life teemed beneath me. I kicked slowly around an imposing coral that formed a boulder as big as I was tall. I had read of these ancient giants of the species *Porites*, and it was humbling to see her sitting here in the glory of this setting. How long had she watched over this piece of sea and what had she lived through in her long life? Alicia swam near me, and I gestured at this coral's immensity, at my awe of her greatness. She nodded, and we swam together along the reef. This felt un-like a restoration; it had taken on a life of its own, become its own reality.

Suddenly, I was torn from my reverie by a sound, abrupt and loud and scary. I looked imploringly at Alicia. She gestured with her hands forming expanding spheres. We had heard a bomb exploding, a reef somewhere within earshot being reduced to rubble.

21

Reef Stars

At dinner later that night at a Chinese restaurant called Surya Super Crab, another place you could never identify from the outside unless you already knew where you were going, Noel ordered enough delectables from the sea to completely fill the turntable in the center of the table: spicy chili crab, black pepper crab, and Singapore crab, along with some of the most massive prawns I'd ever seen. I'm mostly a vegetarian and usually avoid eating farmed animals and nearly all fish because of the toll they take on the environment, but wild-caught animals low on the food chain like crustaceans are one of my exceptions. The meal was a delight.

The talk, as it does with longtime scuba divers, turned to scary dive moments. Noel and Richard took turns recounting tales of running out of air underwater and being forced to grab for a dive buddy's regulator, and of getting the bends, a painful and potentially fatal problem that occurs when the compressed gas a diver breathes at depth forms bubbles that expand inside the body upon ascent. After a few rounds of Bintang, the local beer, the conversation spun again, and we began discussing mistakes that Noel had learned from in his years building reefs.

"At first, we didn't know about the damselfish," Noel said. A few months after Mars installed its first restoration, he had invited a fish biologist to visit the restoration. She'd taken one look at the algae-covered rebar structures and said, "You've got a damselfish problem."

More than seventy species of damselfish swim in the seas. Most look like a two-dimensional cartoon drawing of a fish: flat, with a rounded shape and a comb of fins on top and bottom. They have expressive faces, often with wide eyes and small mouths. They're a problem for coral because they are also master gardeners. These fish cultivate an algal farm a few square yards in size, and they patrol the area against infringing creatures, usually aggressively. When I was diving, rather than retreating as I passed, damselfish darted toward me with their dorsal fins raised to make them look bigger and scarier. Because algae don't have to go to the trouble of building a calcium carbonate skeleton like coral does, it can spread along surfaces much faster than coral. And once it's taken hold, the coral can't grow over it. Add in the damselfish nipping away at the corals to open up more space for their algae farms, and soon the restoration becomes a mossy mess. Now, before installing a new restoration, Noel and Saipul remove the damselfish from the area to give the coral a chance against the algae.

"The other thing we learned was that we did the first installation in the wrong place," Noel said. "We didn't know that we needed to build the spiders up against an already healthy reef." That encourages herbivorous fish that aren't farmers to swim into the restoration and keep the algal growth down. "If there's open rubble with no cover between the installation and the healthy reef, we can't get the natural algae predators to swim out and help us. We have to give them continuous cover."

This critical role that herbivores play in the transition from a new installation to a healthy reef was one that Frank Mars had also mentioned to me. He had teamed up with marine acoustics researchers from the University of Exeter in England who were studying how reef sounds could encourage immigration into the restoration. Earlier in the day, when we'd traveled between restoration sites, we'd bumped into the acoustics team and talked with them briefly as our boats bobbed up and down next to each other.

Tim Gordon headed up the project. He said he had placed underwater microphones on healthy reefs and unhealthy reefs alike and discovered a massive difference in the sounds of each. Far from being a silent sea, the reef is full of noises of fish, urchins, and shrimp chewing and grunting, flapping and scuffing against one another. When an environmental catastrophe occurs, the songs of the reef reflect that change immediately. On the Great Barrier Reef, Tim heard a 75 percent reduction in sound following a bleaching. "When the coral dies, the fish and shrimp that make noise are gone. They are like the birds in the forest without the trees."

Later, when I visited another coral restoration site on the north coast of Bali, I would hear it for myself. There, I met Alice Eldridge, an acoustic ecologist with a background in computer science and music. She had strung half a dozen underwater microphones around a reef, and she handed me a pair of headphones so I could listen to what she'd captured. Most noticeable were sizzling bacon sounds of what she thought were snapping shrimp claws but probably included some yet to be identified crackles. Underneath it were lower sounds, purrs, groans, gulps, and kisses. Switching the file, she said, "Let me play this for you. Maybe you can help me identify it. It happens at half past six every evening." On this recording, she'd done some processing to remove the higher-frequency crackles on top so that the lower noises were more audible. I heard a haunting sound, a low round tone that had the quality of the "ohhh-ommm" from the end of a yoga class. Alice said, "It sounds like a mammal, doesn't it? But there aren't thought to be marine mammals around here. And it lasts for a half hour or so every night."

I just shook my head in mystification. I couldn't fathom what could make that sound.

Alice asked me to listen to one more recording. It was the sounds of the reef, but sped up several-fold. I smiled in amazement. Rather than alien and unfamiliar noises, it sounded exactly like the twittering and chirping of birds in a forest.

At the Mars restorations, Tim Gordon and his team were attempting to use the sounds of a healthy reef to reverse engineer a degraded soundscape. They were installing speakers on the edges of newly restored reefs and playing the sounds of health in hopes that fish would respond. "It's called acoustic enrichment," Tim said. His results showed that it worked. Twice as many fish were attracted to restorations where he broadcast the snaps, whistles, and crackles of health as in sites that were left alone.

Back at the restaurant, Noel signaled the waiter that yes, we'd like another round of Bintang. I considered all that Mars had learned, not just about reinforcing reefs with rebar structures, but about using the ecosystem to their advantage—removing the damselfish that decreased their chances of success and luring in the herbivores with sea sounds to accelerate restoration. I asked, "So, what's next? How do you scale up the Mars methods?"

Alicia said that Mars was holding workshops for the national marine parks around Indonesia. They had plans to install a restoration with a group in Mexico in the next several months and on the Great Barrier Reef in the next year. She said that as with such forms of expansion, there was talk of rebranding the rebar structures that formed the support for the corals and the newly stable reef. So far, the Mars team had been calling the structures reef spiders, or just spiders, because of their multi-legged form. The term also referred to the web that the structures formed underwater once they were installed. But the team was considering calling the structures something that had more of a connection to coral. One name that had been pitched was Reef Stars.

Noel was not inclined to rebrand. "They are already called *laba laba*," he said, using the Indonesian word for spider. "If you rebrand you lose all that clout with the government."

Richard, whose background was in branding, said, "I love the name 'reef stars.' When we were diving on the restoration, it was impossible not to see stars everywhere." He pointed out that as the coral grew and filled the long arms of the rebar, the central spokes were the last to fill in. The final glimmers of reconstruction were the six-pointed rebar centers surrounded by healthy coral. "It just jumps out at you," Richard said.

The waiter arrived with another tray of beer, which he set on the table. As I poured a glass for myself, I realized that I had learned two Indonesian words. We'd been talking about *laba laba* all day. But the beer bottle label caught my eye, a red star on a white background. The word *bintang*, I suddenly realized, means "star."

22

Galaxy of Potential

The next morning, we repeated the previous day's routine: waking to the 4:30 call of the muezzin; guzzling cups of weak coffee; meeting Noel, Saipul, and the dive team at the port; and then boating out to Bontasua. But that day, we would actually build a restoration. It would be a small one by Mars standards: just one hundred spiders, or reef stars, depending on the branding. But I would get to see exactly how it was done.

Approaching Bontasua, we saw an armada of table-sized foam floats moored in waist-high water. Three or four men wearing long-sleeved shirts and gloves, as well as traditional cone-shaped straw hats, surrounded each one. Saipul said Mars paid slightly more than the men would earn for a day's fishing to help with the restoration. Just before I slid into the water, Saipul placed a pointed straw hat on my head too to protect me from the equatorial sun. I bellied up to a float, which held two rebar structures, some fragments of coral, a pile of zip ties, a few pairs of scissors, and a small waste bin. I watched the men tie coral onto the supports and, once I thought I understood what to do, I reached for a coral and a zip tie and started to fasten it along an arm of rebar. A man to my left shook his head at me. The coral needed to be placed properly. I'd started on a section that was supposed to be left open. I tied a coral in the right spot, then moved on to the next piece. But no, I was corrected again. Each coral needed to be tied with two zip ties, not one. The coral needed to be secure, free of wiggles, so that it could attach its skeleton onto the structure. And the loose end of the zip tie needed to be trimmed short, so as not to act as a surface for algae. The trimmed ends of the zip tie needed to go into the wastebasket and not drift into the water. Just as soon as a spider had all fifteen pieces of coral attached, it needed to be submerged in the sea. All the details mattered.

I was nowhere near as fast as the men, but I started to get into the swing of tying on the corals. The number of times I was corrected slowed, and the men started to talk among themselves. As in any group, a class clown was cracking everyone else up at one of the floats to my left. After half the reef stars were ready, Saipul set up a sort of bucket brigade, passing coral-laden reef stars underwater along a chain of people, and then onto the boat in a loose pile. I leaned down, putting my face in the water to grab a spider, but Saipul pulled my shoulder back and said, "Use your foot to lift it." I remembered that there was no sanitation on the island, and we were fairly close to shore.

The clock was ticking, Noel reminded us, as we climbed aboard the boat for the twenty-minute ride to the restoration site. He continuously sprayed the corals with seawater as we traveled. Once there, we geared up in order to unload the reef stars, diving with snorkels to bring them to the seafloor. Speed was critical. We set the rebar structures down wherever we could and then returned to the boat for another and another without stopping. Only once all the corals were safely underwater could we take a break.

When we returned to the island to collect the second group of reef stars, the coral tying was complete. The men were still standing in the water around the floats, but now they'd been converted to coffee break tables. Thermoses and cups were scattered about, as well as a bottle of hot sauce and bowls just emptied from a late breakfast.

After we'd transported the second set of coral-laden spiders to the restoration site, we put on scuba gear and began to build. Alicia and I teamed up. She showed me how to move large pieces of rubble, creating stable footing for each reef star. We wedged the legs of adjacent stars together so that they crossed each other near the base. I understood the reason for the angle of the legs. Rather than a 90-degree bend in the knee, the wider 110-degree angle allowed for two legs to cross and interlock. The crossed legs were zip-tied together in a precise manner. No stars could slip from the galaxy we were forming. The hexagon shape was important, as it allowed stars to fit snugly around pieces of living coral and butt up against the wavy edges of the living reef. Those edges were then staked in place with metal rods. I hadn't brought quite enough lead weight with me to counterbalance my buoyancy. Clinging to the rebar with my arms as my legs canted upward, I basically ended up in a handstand. But as much as I pulled upward, even as the network was under construction, it held solid to the seafloor. Mid–zip tie, I heard a sound, abrupt and dramatic off in the distance. I looked at Alicia, who made the hand motion of a bomb again. I'd hear it once more before we finished the build.

Because of my buoyancy problems, I used up the air in my tank more quickly than the other divers, and I had to return to the surface before the arranging, staking, and tying was completely finished. But before I did, I scanned the seafloor. What just an hour earlier had been a wasteland of rubble now flickered between two versions, like an optical illusion. In an instant, the network looked like a honeycomb of hexagons, and I imagined the latent reef as the hive of activity it would grow into over the next months and years. A second later, I focused instead on the interlocked stars, and I imagined a constellation of coral and a galaxy of potential.

Back on the boat after the restoration was complete, the boat driver turned up the music. The hit summer song, "Despacito," sung by a singer from Puerto Rico, another coral-encrusted island halfway around the world, blared out of the speaker as we packed up. The tune changed and Right Said Fred was singing "I'm too sexy for my shirt," followed by Imagine Dragons wailing about feeling the "Thunder," both of which made me smile, thinking back on how the male divers had turned their backs out of respect while we women put on our wetsuits. Nonetheless, by the time we bounced our way back toward Makassar with the sun just past its apex, all our voices were singing together, "Young, dumb . . . young, young, dumb, and broke."

STEPHANIE HAD STAYED UP LATE and skipped dinner with the rest of us two nights in a row. It had taken her that long to stitch together the 360-degree images of the reef that she and Richard had been shooting. But she wanted to have it done in time for our final trip to Bontasua. As before, we motored across the flat turquoise waters and pulled up to the long dock. We climbed the tire ladder and trundled past the little corrugated metal–roofed shop, the flowering trees, and the waddling ducks. We slipped our shoes off and entered the meeting hall. Again, the four women who served as representatives of the government met us. With Saipul as a translator, Richard thanked the villagers for allowing us to work on the reef and said he had imagery he wanted to share. He handed out several sets of virtual reality goggles and encouraged everyone to take a look. There was collective hesitation. Then Noel took one of the viewers and stared into it. His face transformed like that of a little boy. He turned and twisted, pointing the viewer up toward the ceiling and down toward the ground. He pointed at fish he saw, at divers swimming through the image, at various coral. There's no way not to look foolish staring into a VR viewer, and in that foolishness, the ice was broken.

Noel handed the viewer off to one of the male officials, who performed the same choreography. The viewers reached the women, who responded in exactly the same way, pointing, twisting, looking up and down. Around the room the viewers went, everyone's face growing smilier as the virtual reef was passed from person to person, a world that existed in reality just steps from where we were standing. It crossed my mind that despite our technological advances, the group of us standing together in this room could never connect in such an immediate way in that reality beneath the sea. Masked and fitted with breathing devices, our smiles are limited to our eyes and our conversations to gestures. We can't speak. We can't laugh. The sea remains a buffered place for us humans.

The woman I'd noticed earlier, with the gold-rimmed glasses, seemed to spend more time looking at the virtual world than anyone else. After a while, the little boy who had been wearing the Spider-Man shirt two days earlier scrambled onto her lap. Today, he sported a blue race-car shirt. The woman put the viewer to his eyes. Like everyone else, he intuitively looked up and down, twisted and turned. Trying to make sense of it, he stuck his little foot out and stared down in its direction, testing if it, too, would show up in the world of coral he was visiting. Like all of us, he was trying to parse reality from illusion.

We'd planned to get to Bontasua by eleven a.m., so the men would be free to go to prayers at noon. But I looked over at the clock and saw that the time had trickled to 12:05 and then 12:15. No one was leaving. In fact, a spontaneous meeting had taken place between the Mars team and the people of Bontasua, who were sitting together in a circle of plastic chairs. Afterward, Noel said the conversation was about expanding the restoration. The villagers pointed out that a large boulder was in the center of the area where Noel proposed building. It was a hazard for boats, and the men thought they should put a light on it, so no one would hit it.

"It was the first time they've asked for anything," Noel said. But that wasn't the end of it, he added with surprise. They also asked for English lessons. If visitors would be coming to see the restoration, they wanted to speak to us directly. Bringing the virtual world of the coral into the meeting room seemed to have catalyzed a shift in the conversation, Noel said.

Richard, who saw the reef as its own best advertising, said, "It's the power of imagery."

Perhaps. Or perhaps the shift had already started. Earlier, Noel said he'd hoped we would get the chance to talk to the head of the island, the *kepala desa*. The last time Noel had brought some guests to Bontasua, they had asked the *kepala desa*, "What do you think about this restoration?"

He answered, "The big difference is that before the restoration, none of us talked about the reef. It wasn't a conversation we had. Now, we are talking about it."

BEFORE WE LEFT BONTASUA FOR THE LAST TIME, Stephanie and Richard wanted to take drone footage of the island and its reefs. While they filmed, I decided to wait on the dock where some local kids had gathered. I sat on a wooden bench and pulled out my notebook to take some notes. A few girls perched shyly on the far side of the bench from me. One of them, her dark hair pulled back in a messy ponytail and golden kittens dangling from her ears, seemed like she wanted to meet me. I handed her my pen and paper, urging her to draw something. But she shrank away. I drew a cartoonish cat on the page and showed it to her. She smiled slightly and scooted half a scoot closer to me. Emboldened by their friend, a few more girls joined us on the bench. I drew a few hearts. I wrote my name. They refused to mark the pages of my notebook.

After a while, adults began to assemble on the dock, and soon a low, long sloop piled with all manner of goods arrived. This ferry supplied everything necessary for survival on the island. Three crew people clambered from package to package, lifting those that filled orders up to people on the dock. Jerry can after jerry can, worn and dented, heavy with fresh water needed for cooking and washing, were lofted upward. Saipul told me that though the water was fresh, it wasn't clean, and it would still need to be boiled before anyone could drink it. Then came metal canisters of propane, then some building supplies, and finally groceries. An older woman grabbed a clump of greens, already wilting in the hot air. She split them with a neighbor and retied the stems of her portion. Once all the supplies were off-loaded and the groceries divvied up, the ferry pulled away. The women balanced their packages on their heads and walked toward their homes. The men piled goods on wheelbarrows, which they pushed up the pier. A major problem for the people on the island is that there is no comparable ferry in the opposite direction. The government doesn't provide a sanitation collection system. All of the cartons and containers that were just unloaded would become a disposal problem. The people would be forced to live with their garbage or, if not, to burn it, which was forbidden. Or it would be taken away by the tides.

I was left alone with the girls again. We sat in curious silence, not knowing how to cross the many divides between us: language, culture, and age. Grasping for connection, I thought of a kids' song with hand movements, and began walking my fingers and thumb against each other, singing, "The itsy bitsy spider went up the waterspout."

Foolishness broke the ice a second time that day. The girls overcame their shyness and began walking their own fingers and thumbs, crawling up their own waterspouts. And just then, I realized the song was about spiders. So together we sang of *laba laba*, words familiar to them and the tune familiar to me, over and over, like the many legs of the creature itself.

MY LAST AFTERNOON ON SULAWESI, Alicia and I took a walk through town, weaving between the open sewer, root-riddled sidewalks, and buzzing mopeds. Alicia asked me if I had any remaining questions.

"I guess I still have one big one," I answered. "I still don't really understand why Frank Mars did this."

"I don't know if there's one good answer," she said. "He's told me that when he was a kid he went to Lizard Island in Australia with his family. He dove on the coral there and ever since had a passion for the ocean."

"Yeah, but lots of people have a passion for the ocean. It's another thing to spend fifteen years developing a restoration process and continually funding a team to restore hectares of reef in Sulawesi."

"Right, and there's a bit of leaving a legacy," she said.

"That makes sense," I agreed.

Alicia laughed. "He's also competitive. He likes to win. He wanted to figure out how to do reef restoration and wasn't willing to fail."

That was actually the best answer I'd heard. By 2021, Mars had planted over 280,000 corals in an area just under 4 hectares, making it one of the largest restoration projects in the world, if not the largest. Additional de-ployments were in the works in Bali, in all the Indonesian national marine parks, and in Mexico. A pilot project in Australia was approved and by 2021 planting was under way. In 2021, Mars, through its cat food brand Sheba, built a reef in the thirty-year-old rubble field near Bontasua. Drone footage from the sky above showed a 150-foot-wide network of rebar stars and coral. In four fifty-foot letters, it spelled out the word HOPE. One day, hopefully, the reef stars will be so overgrown with coral that the word will disappear.

PART V

BALI

23

The Coral Cloud

While coral numbers are on the decline in the seas, their abundances are flourishing right here on land—in aquarium-sized reservoirs. More than two million aquarium hobbyists tend and grow living coral in their dining rooms, dens, and basements. Two decades ago, the last time a UN assessment was made of the marine aquarium hobby, it was valued at as much as $300 million per year and included millions of people around the world. While it's hard to tease out what proportion of that comes from the sales of coral, tens of thousands of buyers meet up at annual trade shows with amalgamated names like "Reef-a-palooza" and "Fragstravaganza" and at online auctions where corals are sold in rapid-fire allotments over the course of a weekend. Coral reef scientists have been slow to tap into this distributed "coral cloud," but they are beginning to recognize its value, not only for the genetic diversity but, perhaps even more important, for the expertise in coral breeding and husbandry that exists there.

My foray into the world of aquarium corals began one Saturday in Austin when I received a text message from a neighbor. Richard Gorelick and his son, Noah, maintain a stunning coral aquarium in their living room, and Richard's text said that later that morning he was headed to the Austin Reef Club frag swap in the gym of a church not far from my house. Fragments of coral, or frags as they are called in the hobby, would be traded, bought, and sold. The announcement promised a dozen exhibitors with tantalizing names like Fishworks, Jolt, Pod Your Reef, iJam, and the very sexy-sounding Obsessive Coral Desires. I texted back a thumbs-up emoji.

Even though it was midmorning, walking into the church gym felt like Saturday night at a dance club on Austin's famous Sixth Street. The room was dark, save for a deep blue light cast upward on the high ceiling and reflecting off the polished wooden basketball floor. Ultraviolet bulbs illuminated each of the vendors' shallow tanks, which formed two rows along the sidelines of the court. A father and son sat on folding chairs behind the first exhibitor table and greeted me as I slowed to peer into the bottom of their shallow tank. It was studded with an array of corals that looked like faceted gems resting in a black velvet jewelry box. Using a fiber-optic pointer, the duo gestured to the different specimens, each of which was attached to a small ceramic stub that looked like a fat golf tee and rested in holes in a plastic rack. They were the same stubs used by coral scientists, who borrowed tools of the trade from the hobbyists. I was particularly entranced by the zoanthid corals and their showy bull's-eye-patterned polyps: deep red centers ringed in tangerine, and a fringe of indigo tentacles; sorbet green centers surrounded by circles of yellow, then green and yellow tentacles like the petals of a daisy; lemon-yellow centers encircled by indigo, pale purple, midnight blue, and a flash of green on the tentacle tips. The combinations were infinite, dizzying.

The bright colors of the zoanthids are not just beautiful; they are a warning. Zoanthids are corals, but they don't make a protective skeleton like the reef-building corals I've been focusing on. They belong to a sister group, collectively called soft corals, that don't have the protection of hard armor. Instead, zoanthids contain the chemical palytoxin, which is considered one of the two most poisonous nonprotein substances ever discovered. Scientists studying venoms rate toxicity using an LD score, which stands for lethal dose. It is the amount it takes to kill half the cells exposed to it. The smaller the LD score, the more potent. The LD of the venom of a black widow spider is 0.9 mg/kg; a Mohave rattlesnake is 0.03 mg/kg. Zoanthid palytoxin is two hundred times more potent at 0.00015 mg/kg. According to the Marine Aquarium Societies of North

America, there are no best practices for handling zoanthids, but they strongly advise wearing goggles, long gloves, and a face mask if you do.

Like hard corals, soft corals are holobionts. Recent evidence suggests that symbiotic algae from the genus *Ostreopsis* manufacture palytoxin and then pass it off to the zoanthid, which concentrates it. Or it might be that symbiotic bacteria produce precursors for the toxin, which the algae then finishes processing and passes to the animal host. What's so interesting is that the roles of the symbiont and the host echo each other in soft and hard coral: Stony corals gather sugar from their algal symbionts for the energy to build their protective skeleton; soft corals gather palytoxin from their symbionts for protection, too. "Soft corals have no defense," one coral aquarist said to me, referring to the fact that they have no physical armor. "It's just chemical warfare for them."

In the disco light of the church gym, I slid to the next table, which held a larger tank with an even larger selection of glowing gemlike coral. A man standing next to me decided to buy a nubbin as bright orange as a traffic cone. The dealer, wearing a gray T-shirt with a logo of a branching coral and his website, www.ijamcoral.com, across his chest, filled a small plastic bag with a few inches of water. He dropped the nubbin in and, reflecting off the plastic, it lit up the entire bag.

As I stood ogling the many corals in the tank in front of me, my neighbor Richard identified the cnidarian menagerie. It was hard for me to get a grip on the names that the aquarium community used, which have only the loosest connections to scientific names. Like corals gathered for European museums in centuries past, these corals had no country of origin attached either. They were simply named for their looks. Jack-o-Lantern Leptosteris were orange with yellow eyelike polyps; Mumbo Gumbo Jumbo Goniastrea displayed a sunset of oranges, pinks, and greens; Nuclear Meltdown Lepto was seared in radioactive green; Papa Smurf Acro sprouted pale blue branches; Crocodile Hunter polyps looked like blue scales piped in orange and purple—you get the picture. The most expensive frags sell for as much as $600 for a piece about an inch tall.

Just how does a highlighter-orange coral find its way to a church gym in Austin? The answer is complicated. Data on the international patterns and routes of coral trade is poorly captured and fraught with inconsistencies. Prior to 1980, coral entered the market mostly as dried skeletons, or curios, like the kind I remember resting on a bookshelf in my grandmother's house. Those corals were probably harvested from remote locations in the Coral Triangle and the South Pacific in large volumes, dried, and then sold to tourist shops like the one in a resort town in Mexico where my grandmother purchased hers. Around 1980, scientists began noticing the degradation in coral reefs, and while the problems that still plague reefs today—destructive fishing practices, pollution, and climate change—were taking their toll, much blame was focused on the coral harvest. So, in 1985, the coral trade was put under the jurisdiction of the Convention on International Trade in Endangered Species (CITES). Exports of bulk coral slowed because now coral was required to be classified down to the genus level, something that can be tricky even for experts. Without tissue, coral skeletons for the curio trade were even harder to identify. The additional CITES requirements pushed exporters away from selling coral curios to trading in live coral, which could be sold at higher prices per piece and justified the extra bureaucratic steps. This subsequently created a market for collectors in remote areas, often artisanal fishermen who could bolster their income by harvesting wild coral. According to the 2003 United Nations report on the coral trade, the largest coral exporter was Indonesia at 71 percent, followed by Fiji at 18 percent, and then the Solomon Islands and Tonga at less than 5 percent each. In the years since that report was issued, Australia has become an important exporter of wild coral as well. In terms of imports, the United States accounted for 73 percent and the European Union 14 percent, with Japan, China, Canada, and South Korea as smaller markets.

When the trade follows legal pathways, which isn't always the case, small-scale collectors pass their finds on to middlemen who deal with obtaining the CITES permissions and then ship the coral around the world. But the CITES system is notoriously flawed, and illegally harvested wild coral can be slipped into the system, relabeled, and redistributed along with the legal specimens. The pretty piece of highlighter-orange coral in a tank in a church gymnasium in Texas could have been born and grown up on a faraway reef in Sulawesi, been plucked by hand by an artisanal fisherman who had an agreement with an exporter, traveled through an airport in Jakarta with CITES permits, cleared customs in L.A., been shipped by mail to an aquarist in Dallas and driven down I-35 to Austin. Or it could have been taken from that same reef by a smuggler, shipped to Singapore, relabeled and mixed in with legally harvested coral, and then sent on to Dallas. There's no way of telling.

By the mid-1990s, the growth in the supply of live coral helped change aquarium fashion. Rather than fish swimming against a poster of a coral reef taped to the back of an aquarium (like I remember from my childhood), hobbyists began to create elaborate miniature ecosystems inside their tanks. They craved particular types of coral that could play a role, either functional or aesthetic, in a tank environment. Andrew Rhyne, who teaches marine science at Roger Williams University in Rhode Island and is an avid aquarist, led a study of two decades of CITES permits on imports to the United States. He found that the live coral market grew about 10 percent per year from 1990 to 2006, feeding this craving for ecosystem-based aquaria. But then it fell by about 10 percent annually. By phone, Rhyne told me the decline was partially attributable to a global recession when consumers slowed their purchases of nonessentials, but also to the emergence of coral farmers and cultured corals.

Unlike wild harvesting, in which coral is taken from the reef, farming has the potential to be sustainable. A farmer finds a particularly beautiful specimen, a coral with a special color or a shape that will appeal to an aquarist. Rather than sell it to a middleman, the farmer replants it in an undersea ranch or in a land-based aquarium farm. This "mother" coral will not be sold, but instead its limbs or edges will be fragmented and grown for a few months to a year, depending on how fast the animal grows. Once it reaches a large enough size, the fragments will be sold. While not all species of coral popular in the aquarium trade can be farmed, some lend themselves well to domestication, especially fast-growing branching coral in the genus *Acropora*. By 2010, 26 percent of *Acropora* were classified in CITES codes as farmed, bred, or ranched. So it's also possible that the highlighter-orange coral in the church gym wasn't plucked from the wild off a remote reef, but rather it and its fragmented clones grew up together on an underwater metal rack in an industrial bay near an international airport in Bali.

In 1997, feeling pressure from environmental groups, the government of Indonesia set up quotas for coral exports, hoping to minimize over-harvesting of wild coral. The European Union questioned how the Indonesian government came up with those quotas, claiming they were too large, and banned the import of about a dozen species. The moratorium spread to the Philippines and to Fiji, even reaching Hawai`i. The following years have seen starts and stops of stony coral trade to and from various countries. The reasons given for the repeated shutdowns are largely a lack of information about coral identification and life cycles. Like so much else in the sea, funding for that research hasn't materialized.

A FEW MONTHS BEFORE I PLANNED to travel to Indonesia, the country implemented a sudden, new ban on coral exports. *The Jakarta Post* reported that the shutdown was not a result of a coral corruption ring or violations by coral exporters. Rather, its genesis was a disagreement between two government agencies over discrepancies in CITES certification and monitoring. With more than twelve thousand people involved in Indonesia's coral industry, collectors, farmers, and exporters were all worried about the economic toll if the moratorium dragged on too long.

I started visiting one of the more active aquarium websites, Reef Builders, to follow the news on the ban. Early on, Reef Builders said the freeze was just a technical rub between two agencies, and they expected it to be resolved "within days." As the ban dragged on unresolved for almost a year, Reef Builders filed a post titled "The Indo Ban Is Still On, but Corals Are Growing Out Nicely," with photographs of coral crops in

Bali overflowing their metal racks like cascading bridal bouquets. One of the featured farms was called Ocean Gardener, headed up by Vincent Chalias. After completing his master's degree in marine science in his native France, Vincent traveled to Bali to learn about the ornamental fish trade, and never left. He teamed up with a Balinese partner and set up the first Indonesian coral farm, Bali Aquarium, which now includes a dozen outposts throughout the country. By 2014, 80 percent of the farm's exports were cultured.

Vincent frequently wrote articles for Reef Builders and I could hear his love of corals dripping off the page amid otherwise technical descriptions of coral anatomy. He wrote about a single species of particularly beautiful *Euphyllia*, giddy over its rarity and the "delicacy of its skeleton"; about the biological reclassification that included the flowerpot coral with its twelve fat petal-shaped tentacles that make it look like a daisy: "We knew for a long time they were different . . . But from now on, we know why!"

When my plans to visit Sulawesi had firmed up, I reached out to Vincent to let him know I'd be nearby. He graciously invited me to visit his coral farms on the east coast of Bali between my trip to Sulawesi and when Keith and the kids would arrive for our family vacation. Although Stephanie had obligations in the United States, Richard had scheduled visits to a few coral restoration projects around Bali, in part to see if he might be able to find a site for his client to make that $10,000 donation. And so Richard decided to join me on my visit to Vincent's farm.

24

The Coral Farm

As Richard and I waited for Vincent to pick us up at our hotel, my phone beeped with WhatsApp messages.

"Hey Juli, this is Vincent, I'm on the way"

"Still stuck in Denpasar traffic though"

"Big ceremony on the way, we will lose 15-20 mn"

"Completely stuck, not moving . . . not sure how long it will take . . ."

An hour later, Vincent blasted out of his van, tall and thin, with an angular face and close-cropped hair. He apologized, explaining that Bali was just about to be swept up in a week of celebrations for Galungan, a business-stopping holiday similar to Christmas, and preparations were gumming up traffic. We hopped into Vincent's minibus to drive the short distance to his farm. "I don't drive in Bali," he said dismissively in his French accent as he dodged motorized bikes and flew around curves. "It makes me crazy. I try to stay sane."

I wasn't sure I believed him.

On the streets, evidence of the coming holiday was everywhere. Soaring bamboo poles, called *penjor*, perched next to every mailbox. They were lavishly decorated with loops of banana and coconut leaves, like the mane of a giraffe done up in hot rollers, and hung with strands of rice, fruit, and gold and white flags. The very tip of each pole was finished with a long chain of banana leaves holding a large circular ornament, representing the cosmic circle of life. It was positioned to dangle just above the height of passing trucks. In its curved shape, I learned, the *penjor* took the form of the great mountain Agung, where the Balinese gods reside. Made from only natural materials, the *penjor* represented the importance of the relationship between man and nature. Collectively, the many *penjor* became an elegant and festive canopy over the streets. Passing beneath them reminded me of swimming through the swaying fronds of a kelp forest.

As we drove, Vincent talked about the effect of the coral export moratorium on his business. In order to support his staff and farming efforts, he had created a nonprofit arm of his company, which focused on restoration and education. Working with dive shops, he'd developed a half-day course on coral biology for tourists. We were headed to the newly built classroom. Vincent parked on the edge of a cliff near a Balinese temple, and we followed him down a narrow path, which had a breathtaking view of the bay where one of his farms was located. Typical Indonesian fishing boats with single outriggers bobbed on their anchor lines, and green mountains rose out of the sea in the distance. On one side of the classroom was a large room stocked with masks, snorkels, and booties tourists could use when they signed up for the class. On the other side, a PowerPoint-like presentation was printed on large sheets of weatherproof canvas. I flipped through the pages of the presentation. One showed the currents of the Indonesian Through Flow that made Indonesia such a great place for coral to grow. Another contained beautiful photos of soft corals like the zoanthids I'd seen at the coral swap meet in Austin. I paused on a page dedicated to the genus *Acropora* showing its center of diversity in the pocket of northern Sulawesi, just on the other side of the island from the Mars restoration site, a place that acts like a catcher's mitt for whatever coral genes the great Pacific Ocean beyond delivers. In the pocket of that mitt, time and gene mutations have created the most diverse genus in the coral world. Of the 800 or so species of hard coral, Vincent said when he noticed me looking at the page, about 20 percent were members of just one genus, *Acropora*. Within that *Acropora* genus, 129 different species were found in Indonesia.

I was reminded of my first conversation with Misha Matz. He had pulled a skeleton of *Acropora* off a shelf in his office at the University of

Texas. He'd pointed at the polyp at the end of a branch and said, "This is the innovation that changed everything. *Acropora* invented the axial polyp. And it allowed them to grow fast." Like having eyes in front of our faces, having a polyp at the very tip of a branch gave the *Acropora* coral a direction to move. *Acropora* became the weeds of the reef, not just in the Indo-Pacific but in the Caribbean as well. Those tips are so optimized for growth that they grow faster than the symbiotic algae can colonize them and turn them brownish. That's why a field of branching coral looks polka-dotted in white, and all the more attractive to our eye. But it's not there to impress us. It's a side effect of speed.

With our masks in hand, Richard and I followed Vincent down another narrow path to the water. Fins weren't allowed near the farm—too easy to break the merchandise—so we awkwardly walked and kicked in our booties to the middle of the bay. The farm itself was a group of rectangular metal racks set about a foot off the sandy bottom. Each rack held many clones of a single mother coral. Together, the colors were more brilliant than any wild reef: lime greens, lavenders, deep purples, teals, aquamarine, burnt orange. One was green down at the base, but halfway to the end of each branch it switched to watermelon pink.

We'd timed the visit during low tide so we could stand and talk in chest-deep water. Vincent picked a lovely orange *Acropora* colony the diameter of a basketball from one rack, holding it like a bouquet by a concrete base. He pointed to joints where he would slice off branches and mount them on their own concrete bases before selling them. A single branch the size of your pinkie would sell to an exporter for about $8. By the time it got to a retailer in the United States, it could sell for as much as $80. The tenfold discrepancy was one of the problems with the industry, Vincent said, a consequence of many layers of middlemen between farmers and retailers. Eliminating some of the in-between transactions was something he was hoping to accomplish in negotiations with the government as they worked to end the ban. He was also advocating for more traceability of corals, some sort of marker so people could know that they were buying cultured as opposed to wild.

"Is this the original coral you harvested from the wild?" I asked.

"No, I keep that somewhere else." The original mother coral— Vincent held his hands apart—was as big as a small bush; he'd had it a long time. Vincent said that according to his permits, he is allowed to sell only a certain number of fragments based on the number of brooders he's got on the farm, and that's regulated, too. Additionally, he was required to give away 10 percent of his crop for restoration.

Vincent gestured toward the fishing boats at anchor and said that he paid the fishing cooperative that owned the bay for the rental of the space for the farm. In fact, the fishermen technically owned the corals until Vincent sold them. They helped with the maintenance, clearing algae and plucking off coral-eating snails called *Drupula*, and he paid them for that as well. The racks were a little grungy with algae, he pointed out. With exports halted, he hadn't been able to pay for as much maintenance. Over the last year, he'd given away his oversupply of coral for use in restorations. When Richard and I visited coral restoration sites in the north and west coasts of Bali over the next few days, we'd see the bright purples and greens of Vincent's resplendent corals replanted there.

Vincent motioned for me to follow him. "Come, I will show you the evil algae."

"Algae isn't evil," I said, shaking my head.

"This one is evil!" he insisted. We swam away from the farm to a nearby bit of reef, a field of brownish branches with white tips. Vincent reached down and yanked off a handful. If it hadn't been for the snorkel in my mouth, I would have screamed, "Stop! Don't break the coral!"

Vincent handed the clump to me. I turned it around in my hands. It was the same color, same shape, had the same size branches, same puckers and knobs as *Acropora*. I felt the same disorientation I've felt holding a piece of 3D-printed plastic, something that looked like it should be heavier and harder, but was deceptively light and squishable, like rubber. I wasn't holding a piece of coral at all, but a seaweed masquerading as a coral. Known as mimic algae (*Eucheuma arnoldii*), the species is edible and is harvested commercially for use as a food thickener and stabilizer. It's thought that fish find it tasty too, and so it evolved protection in the camouflage of a hard coral to deceive them. While mimic algae were rare before the 1970s, now they are common. It's possible a decline in herbivorous fish and urchins might be contributing to the uptick. In most places, mimic algae grows in and around *Acropora*, squeezing the coral for space on the reef. I watched as Vincent ripped out handful after handful, clearing the way for coral to return.

IT WAS RUSH HOUR by the time we climbed back in Vincent's van, so Richard and I invited Vincent to join us for a Bintang to wait out the traffic. Still salty, we sat on a restaurant patio on the edge of the sea and watched the sun set beyond the horizon. As we talked, Vincent wasn't shy about sharing his opinions. Divers: They take unnecessary risks. Corruption: Why do I pay to have so many permits hanging on my wall when the other guys don't? The Fisheries Department: Just shut us down without even taking the time to understand what we do. Fish: They get in the way of seeing coral. The list of grievances also included tipping, English food, algae, institutional racism, and Orlando, Florida.

"You even have a problem with Orlando, Florida?" I asked.

Vincent had been invited to give a keynote at the Marine Aquarium Conference of North America meeting, one of the industry's largest. It would be in Orlando at the end of the summer.

"What are you going to talk about?" I asked.

"I'm going to say what I see. That the aquarium industry is going to be more and more in the spotlight. That the government isn't doing the regulation it should be doing. And so we are going to have to replace what they aren't by regulating ourselves."

Despite his outwardly brash demeanor, I was struck by how much Vincent considered coral from their point of view, the ways he tried to understand them at the level of the polyp. He talked about the "slow wars" that corals fight with one another, using long, sweeper tentacles armed with the sturdiest stinging cells to fight back others impinging on their territory. He wondered what made one coral bleach while the coral next to it didn't. He spoke about the fact that some coral brood their offspring while others broadcast them, and how that affects where he'd find them on the reef. He noted how coral in the wild eat anything that comes their way, but coral in tanks are rarely fed and rely more heavily on photosynthesis. "That changes their shape," he said. Corals denied food puff up to capture more sun. Meanwhile, coral in the wild use their shapes hydrodynamically, contorting to create slow-moving eddies that slow down the plankton for easier capture. Vincent talked about how wild corals have the ability to withstand storms, and murky water, and changes in pH, that their world was so much more turbulent and dynamic than the sunny photos of sunny reefs in nature articles led us to believe. "Corals are fascinating," Vincent said, "like a world without end, like a Russian doll."

In January 2020, after almost two years, the Indonesian moratorium on the trade of cultured corals was lifted when the new minister of fisheries, Edhy Prabowo, took office. Critically, Prabowo actually held a meeting with the coral exporters to talk about the mix of wild and cultured corals allowed out of the country. Exports of cultured corals were given the go-ahead, with additional inspections and quarantine periods. Vincent quickly readied three shipments of coral. But just as they left the country, someone was caught trying to slip a shipment of wild coral in with the farmed. Prabowo prosecuted the smuggler and added a thick layer of paperwork to exports. I spoke to Vincent later that month as he zipped around town on a motorbike—so much for never driving. He said the new bureaucracy was so opaque, none of the coral farmers could navigate it. Given the continued lack of income and uncertainty, Vincent had had to lay off forty employees, even as he had "a thousand customers waiting and a hundred thousand corals ready to ship."

MANY MONTHS LATER, at the beginning of the COVID crisis, I sat socially distanced on the back porch of Richard Gorelick, who had first taken me to the coral swap at the church gymnasium, and his son, Noah, who was headed into his junior year of high school. Noah had asked for some help coming up with ideas for a major science project. The topic was, of course, coral.

"What kinds of corals do you have in there?" I asked, looking through the window at his aquarium, which was a centerpiece of the living room. A parrotfish darted across the top.

"Some acros, some chalices, some flowerpots, some hammers and torches."

The names continued to confuse me. "A good place to start would be getting an inventory of what you have, with scientific names," I said.

Noah nodded. "What we need is a way for the hobbyists to identify the coral, like a DNA test," he said. "And then I can use that as a model to get other hobbyists to add their information to build a database."

"Like a 23andMe for coral," Richard brainstormed, referring to the DNA-sequencing company named for our twenty-three pairs of chromosomes.

"It's a great idea," I said, "if it only existed. Also, I wonder how many chromosomes corals have." Later, I learned it would be more like 28andMe. We talked about the continued gap between the scientific world and the world of hobbyists, and the ways in which each could benefit from the other's expertise. Hobbyists would love to use the technologies that the scientists have developed to know what's in their tanks. The scientists could use the coral collective as a rich genetic bank, not to mention their expertise in growing coral. Also, hobbyists are a group of people already financially invested in coral and among the most passionate about corals. This, perhaps, is the greatest connection that the people working on the science and management of reefs haven't yet made with the world of coral aquarists. This shared passion for the beautiful, mysterious, and important world of corals would have so much potential if only the two groups reached out to each other a little more. Like so much with coral, the message is about collaboration.

"Each home aquarium is basically an experiment and yet none of that information gets recorded anywhere," Noah said. "I'm pretty sure if we started keeping track of it in a more scientific way, some really cool stuff would fall out."

I'm pretty sure it would, too.

25

A Prayer for the Sea

I felt the rumble seconds before the word for what was happening formed in my head. Years living in Los Angeles during graduate school etched the feeling in my subconscious. I jumped out of bed and crouched near the floor. The word moved from a thought in my mind to sound waves in my mouth.

"Earthquake!! Earthquake!!" I yelled.

Keith, who had arrived with the kids in Bali just hours before, rolled over groggily. "What?"

I crawled toward the door and flung it open. The morning light had just broken. "Earthquake!!" I ran outside into the dewy grass. Keith and Isy followed, bewildered and sleepy. Ben, too tired to move, stayed put. In our pajamas, we shyly acknowledged the other guests at the small villa where we were staying. When the rumbling stopped and we saw there was no damage, we all slinked back to our rooms.

"Welcome to Bali, situated on the Ring of Fire, where you're constantly reminded that plate tectonics is happening," I said to my family.

"Cringe, Mom. Not now," Isy said, grumpy with jet lag.

I was in what's not so fondly called "science with Mom" mode, and couldn't stop myself. "You know, when Darwin was on the *Beagle* and thinking about coral, he was in a huge earthquake, too."

"Stop," Isy said, stretching the word into three syllables and flopping herself back in bed.

I was so happy to see them I didn't care that they were annoyed with me. We'd planned a couple of days to get over jet lag in the capital city of Denpasar. Then we'd go to Ubud, the cultural heart of Bali, and end in the small town of Tulamben, in the northeast, known for its diving. Just like Isy, Keith and Ben had gotten scuba certified in our chilly Texas lake. The warm waters of Indonesia were going to be very different for them.

Throughout the week, decorations for the coming holiday grew more elaborate. In the many open-air Balinese temples, we saw gold and white cloths draped around altars and gold and white umbrellas erected over statues of gods. Woven baskets of offerings, sweets, and flowers accumulated near every altar and littered the streets. The *penjor*, those soaring bamboo poles swaying over the city streets, became forests. Markets were jam-packed with shoppers and the motorbikes that brought them, everyone stocking up for the coming festivities. We learned that Galungan day is the start of a ten-day period during which the manifestation of the ancestors and gods descend from Mount Agung, the highest peak in Bali. The holiday commemorates the constant struggle between good and evil and the ultimate triumph of good. Tulamben, the dive town where we would spend the last three days of our trip, sits in the shadow of Mount Agung. When we reached it, I looked upward toward its peak, feeling the auspicious power of its presence.

Our first day of scuba diving in Tulamben focused on the centerpiece of its coastline, the wreckage of a cargo ship torpedoed in World War II that ran aground on the beach. A large earthquake in 1963 shifted the vessel, rolling it offshore, settling it at ideal depths for scuba. One end rested about thirty feet below the surface and the other near one hundred feet. As we explored the sunken ship, I was disappointed that the encrusted marine life was mostly soft coral and sponges rather than hard corals. But for Ben, always fascinated by vehicles of any sort, it was a playground. I laughed through my regulator when Isy and I found him taking selfies pretending to turn a rusty steering wheel. As I had during her certification dives, I marveled at how Isy's anxiety and OCD were quieted when she was swaddled in the sea. She was so confident and assured underwater that my worries about her receded. She emerged from every dive happy, ready to descend again as soon as possible.

One morning we woke before sunrise for an early-morning dive. We hoped to catch the sunrise shift: nocturnal fish heading to rest, and day-time foragers waking with the dawn. Soon after we crossed an open sandy area, but before we reached the wreck, we chanced upon a large school of massive bumphead parrotfish, camel-like protrusions from their fore-heads creating unmistakable silhouettes in the pale morning light. We re-mained frozen by their magnificence as they lumbered past for parts unknown.

Our last day in Tulamben was Galungan, and we started our dive in a small grove of Balinese idols placed in shallow water in front of our hotel. We submerged and swam from one statue to the next, inspecting each with the buoyant three-dimensionality that diving allows. We moved on to a coral restoration site we'd been told was nearby. When we arrived, we found rounded concrete structures with holes in them, called fish balls, chained together in groups of two or three. The area was dismal and murky, growing algae to the exclusion of hard coral. It was a disappoint-ment compared to the Mars site, but not unexpected.

During our visits throughout Bali, Richard Vevers and I had repeat-edly seen similar examples of sites that called themselves restorations but fell far short of what we'd seen in Sulawesi. Seeing the Mars project first had given me a false sense that success came easily. Yes, Indonesia was a prime place to restore; it had all the environmental advantages for restora-tion to work. But without the constant research and development, with-out continuous effort and funding, it was not a guaranteed success. "Despite being well-intentioned, they are doing it badly. That's a chal-lenge," Richard had said of the restorations we'd seen. A real part of that challenge was that the groundwork for investment was lacking. Richard had come to Indonesia hoping to find a restoration project to which he could donate that $10,000 on behalf of his client. He left without finding one he felt comfortable endorsing.

After our morning dive on Galungan, a palpable excitement and energy filled the air and mingled with the smell of incense burning in the many temples around town. Motorbikes whizzed beneath the canopies of *penjor*, bearing people dressed in gold and white traditional clothing to and from the homes of their relatives. When we passed him on a path in our hotel, our dive guide was dressed in his own beautiful sarong. He filled Isy's hands with more fruit than she could hold. Just above the shoreline we came upon an altar, draped in gold and strewn with banana leaf offerings, cakes, candy, fruit, and flowers. A young mother, in a lacy white top and a golden cummerbund with flowers behind her ears, added to the collection. Her husband held the hand of their young son. They both wore golden sarongs edged in white, and traditional headpieces, a kind of scarf wrapped around the head and edged in gold. The child's was a miniature version of his father's, but with spiky feathers popping out of the top. I watched as the mother turned toward the ocean and offered her prayers. I followed her gaze, offering my own.

PART VI

DOMINICAN
REPUBLIC

26

The Scale of Tourism

At the Reef Futures meeting in Florida, the Spanish hotel chain Iberostar had caught my attention. Iberostar sponsored the meeting, and Megan Morikawa, a coral genetics scientist who worked for the company, gave one of the keynotes. Megan introduced Iberostar's Wave of Change program, an initiative centered on the hospitality industry and its role as stewards of the sea. There is a business case to be made, she said: A healthy ocean enhances the value of their properties.

Tourism accounts for 14 percent of the Caribbean's economy, more than in any other region of the world. The sector employs 700,000 people and indirectly supports 2.2 million. Many hotels have stepped into the coral restoration world. Sandals Resorts in St. Lucia has partnered with the nonprofit Clear Caribbean and the government to develop a restoration project. Buddy Dive Resort in Bonaire has worked with the government and the nonprofit Reef Renewal to collect and grow staghorn and elkhorn corals, and more than twenty thousand of them have been replanted. In Cozumel, along Mexico's Yucatán, tourists can book a snorkel trip through a restoration site or sign up for scuba classes to learn more about coral restoration. In Jamaica, where the corals were decimated following the Caribbean's sea urchin collapse in the 1970s, local fishermen have pioneered coral gardening and partnered with the Round Hill Hotel near Montego Bay. The most successful restoration project in the Caribbean is Fragments of Hope in Belize, which has innovated techniques to grow and restore branching coral. It offers guests kayak trips through its restored gardens. In the Dominican Republic, Fundación Grupo Punta-cana has been farming coral for more than a decade. Seeking to make a significant addition to these efforts is Iberostar.

Like Mars, Iberostar is a multigenerational family–owned business, controlled by the Fluxà family. Wave of Change was the vision of Gloria Fluxà, the founder's granddaughter. She is also the company's chief sus-tainability officer, as well as co–vice chair. When Gloria and I spoke by video call, she was dressed in elegant black and white; her cropped blond hair formed a stylishly disheveled halo around sea-blue eyes. Gloria stud-ied business at Villanova and Harvard, but said she had struggled to find her place in the family business. A few years ago, while traveling alone, she reflected on how she might bring her passion for the sea to the mis-sion of Iberostar. Returning home, she presented her thoughts to her sister and father, who are the CEO and president, respectively. Eighty percent of Iberostar hotels and resorts front the beach, she pointed out, "and I told them, we can link leisure hospitality to the health of the oceans. And they saw the value that it could add to the identity and how we envision our company ten years down the line."

Gloria incorporated three pillars of marine stewardship into the com-pany's business practices. They committed to becoming free of single-use plastic by 2020, waste free by 2025, and carbon neutral by 2030. Solving the problem of single-use plastic meant removing more than eight hun-dred items from their operations and replacing them with five hundred new ones. It also meant avoiding sending 607 tons of plastic, equivalent to 70 million plastic bottles, to the landfill in 2019 alone. And along the way, more than thirty thousand employees and untold numbers of guests began to rethink the idea of plastic use. The second pillar was responsibly supplying seafood. Either the Marine Stewardship Council or the Aqua-culture Stewardship Council has certified fourteen Iberostar restaurants as sustainable. These restaurants no longer serve any species on the In-ternational Union for the Conservation of Nature's Red List, including many species of grouper and shark, and they abide by the Dominican

Republic's annual lobster closure season, considered critical to supporting the recovery of local populations.

The third and last pillar focused on the health of the coastal waters surrounding their properties, and the first target was coral reefs. Together Gloria and Megan Morikawa assessed the company's properties, looking for a location to begin. The Dominican Republic rose to the top of their list. Its coral suffered significant loss because of hurricanes and disease, yet healthy reefs still existed in places. The country had recently established two very large marine protected areas on the western side of the island near Iberostar's first Caribbean property, built in 1993. There, not only did Iberostar have decades-old relationships it could leverage, it knew the culture. Gloria also had an emotional connection to the resort—it was where she had first learned to scuba dive.

Gloria and Megan reached out to other groups already at work on coral reefs in the country and found a vibrant community that coalesced into the Dominican Coastal Restoration Consortium in 2017. To the north, Fundación Grupo Puntacana, the social and environmental arm of Puntacana Resort & Club, had built a coral research lab and an underwater coral nursery and had started to work on microfragmentation using the methods that Dave Vaughan and Mote Marine Lab pioneered. To the south, the nonprofit organization FUNDEMAR, the name a concatenation of Fundación Dominicana de Estudios Marinos, or Dominican Foundation for Marine Studies, focused on coral reproduction. Iberostar concluded that their most important role would be coral genetics, which also dovetailed with Megan's scientific expertise. At Iberostar's Bávaro Suites resort in Punta Cana, an underused yoga palapa was retrofitted into a sophisticated lab to study local coral's genetics and tolerance to higher temperatures.

Not long after the Reef Futures meeting, I received an announcement from Wave of Change calling for grant proposals to work at their new coral lab. That work could include outreach, such as writing. I applied and was thrilled when I received the award, which provided five days of accommodation at Iberostar's Punta Cana property. I hoped to time my

LIFE ON THE ROCKS

visit to the coral mass spawning, a synchronous release of eggs and sperm into the sea during a few select warm summer nights. I wanted to see what's been described as the sea transformed into a reverse snow globe. To help figure out the optimal dates, FUNDEMAR sent me a spawning-prediction calendar. The best night looked to be August 22, when three, maybe four, different species were predicted to spawn. I planned to arrive two days early, just to be safe.

ALONG WITH MY TRIP to the Dominican Republic, I considered a stop in Bermuda to meet another group working on coral restoration that stood out at the Reef Futures meeting. Gator Halpern and Sam Teicher met at Yale School of Forestry and Environmental Studies. Rather than viewing healthy reefs as adding value to an existing business, as Gloria Fluxà did, they saw coral restoration as its own business opportunity. Their company, called Coral Vita, envisioned an international network of on-land farms where coral could be grown under controlled conditions, sidestepping environmental uncertainties that could be catastrophic in undersea nurseries. The farms would also expose visitors to the potential for coral farming and threats to coral health. The model relied on the microfragmentation and reskinning techniques developed by Dave Vaughan and Mote Marine Lab to grow corals fast and replant them on the reef. Coral Vita's vision had already garnered prestigious recognition, such as a Forbes 30 Under 30 Entrepreneurs listing and a J.M.K. Innovation Prize.

When we spoke by phone in the spring of 2019, cofounder Sam Teicher said their on-land nursery system could reduce the cost of a coral replanted on the reef to $10 or even less. Coral Vita farms located near resorts would provide guests with the opportunity to view and learn about corals without getting in the water. For the more intrepid, underwater restoration packages would provide diving guests the experience of planting new reefs alongside the Coral Vita team. The value of healthy reefs extended beyond the benefit to guests, Coral Vita recognized, providing important storm breaks and fishing grounds. Coral

Vita's vision was ambitious, Sam acknowledged. "We need to rapidly scale up reef restoration, so our thought was, if we can get hotels, governments, reinsurers, coastal property owners, developers, cruise lines, corporate sponsors—you name it—to pay to protect the reefs they depend on, we can create a large-scale financial model to support large-scale restoration." This was the sort of business and financial infrastructure that had been missing in Indonesia when Richard Vevers was looking for a place for his client to make a donation.

Sam drew parallels to what has become a vibrant industry in the land-restoration space. The United Nations estimates that as much as two billion hectares of land are degraded, cleared of the trees and forests that once covered them and with soils that are too poor to support crops. That's an area larger than all of South America. Every year an additional ten million hectares are added to that tally. What's amazing is that restoration of the land is not just possible; it's profitable. South Korea undertook massive land restoration programs, doubling the number of trees in the country. That led to improved air quality and other economic benefits equal to $92 billion, or 10 percent of their gross domestic product. In Niger, farmers cultivated two hundred million trees alongside their croplands, improving agricultural yields and household incomes, not to mention increasing biodiversity and sequestering carbon from the atmosphere.

In the United States, efforts to heal lands, either by returning them to an approximation of their wild condition or converting them back to conditions usable for agriculture, are growing quickly. Land restoration directly employed 126,000 people in 2014, more than other sectors considered key to national economic competitiveness such as logging, coal mining, and steel production. Compared to the oil and gas industry, which supports 5.2 jobs per $1 million invested, restoration supports between 7 and 33 jobs per $1 million invested. The average job in land restoration pays about $50,000 per year and generates about $75,000 in economic output. Its total economic output is $9.5 billion. While these numbers include coastal rehabilitation of wetlands, there is currently no restoration economy for coral reefs.

Coral Vita's vision had impressed investors, too. They had raised a seed round of $2 million in funding. Then they'd created a partnership with the Grand Bahamas Port Authority in Freeport, which provided land for their farm. Cruise ship companies would pay for passengers to visit the farm and potentially adopt or plant coral. Those revenues would then be used to fund expansion. When we spoke, Sam told me that ground-breaking was scheduled to begin in about a month. It wouldn't make sense for me to stop and see the site yet, we agreed. It would be better to wait until the facility was finished.

"This is our proof of concept," Sam said, "and then we want to scale up and put large-scale farms in, hopefully, every country with reefs around the world."

Although the timing was off to pay a visit, I decided to keep Coral Vita on my radar.

27

Corals on Ice

Even after finding out about all the challenges facing Caribbean coral, such as disease, loss of urchins, hurricanes, pollution, overfishing, and climate change, I discovered one more problem that was particularly brutal: a lot of Caribbean corals no longer make babies. What's so crushing is that when reproduction stops, extinction has already happened, even if the corals are still standing. I heard Caribbean coral referred to as both "functionally extinct" and "evolutionary zombies." I couldn't help making the comparisons to my own experiences as I passed through middle age. It was as if the coral populations were entering a kind of system-wide menopause, though their hot flashes were driven by climate change rather than internal hormones. As with my own hesitancy to talk about menopause, I sometimes thought, the fate of the reef has been too often avoided.

In response, scientists working in the Caribbean have taken a two-pronged approach to coral reproduction: capture as much of the genetic diversity that still exists, and enhance the fertility of the corals that remain. When I arrived in the Dominican Republic, Iberostar's lab had been up and running for just a few months. A room-sized filtration system had recently gone online, drawing seawater from a deep brine well in order to avoid pulling in any bacteria or viruses from the nearby reefs that could spread disease. Inside the lab, workbenches lined one wall. On the opposite side, a set of four 1,200-liter tanks that could precisely control the temperature and pH of the seawater were undergoing calibration tests. Megan Morikawa and Gloria Fluxà originally conceived of the lab for genetics experiments in order to identify individuals that might be able to survive future warmer and more acidic seas. However, with SCTLD now discovered on the opposite side of the island, the lab was quickly being requisitioned as an ark of local genetic diversity like what I'd seen in Florida. In shallow tanks, dozens of coral fragments hung from pieces of fishing line. Megan said they were planning to add to their numbers, with hopes to soon house ten different species and over 180 fragments.

Keeping living corals in arks is one way to preserve genetic material. But it is time-consuming, expensive, and worrisome. Each time I passed the Iberostar lab, an aquarist was busy with one of the dozens of constant tasks required to keep the animals alive: cleaning algae, checking water quality parameters, maintaining filters. The lab director, Macarena Blanco, was in the midst of developing a hurricane contingency plan. What if the power cut out? What if the lab had to be evacuated? Where could the coral be safely moved? How long could the coral survive without a backup generator? How long could they survive if they had a backup generator but had to be moved inside? Could they find backup artificial lights? The questions were endless, and nerve-racking.

"Where I come in," the Smithsonian's Mary Hagedorn told me by phone, "is looking into the future." Rather than keeping genetic stocks of corals in flowing tanks of seawater that require constant vigilance, her aim is to keep them frozen, indefinitely. Mary's interest in cryobiology began when she ran across an article titled "Frozen and Alive" in *Scientific American* in 1990. Hundreds of animals, the article said, freeze every winter: caterpillars, barnacles, mussels, snails, even vertebrates like frogs, turtles, and snakes. "While frozen, all these animals show no movement, respiration, heartbeat, or blood circulation. . . . Ice accumulates in all extracellular compartments . . . crystals run under the skin and in between muscles." That last bit is particularly hard to imagine. On the micro level, a growing ice crystal is like a knife that can cut apart cells, spilling their contents everywhere. The way that animals counteract the slicing is by dulling the blade. Some make antifreeze proteins that stick to the first bits of crystal, preventing additional water molecules from attaching to the growing edges. Other animals add special proteins to their blood to control exactly where and when the freezing happens, like guiding the knives into the correct holders.

At the time, Mary worked in fisheries science, and she recognized an immediate use for the technology. Like most coral, many fish spawn only once a year, so fish farmers have a single shot to grow their brood. Failure can be catastrophic. But if cryogenics could provide embryos throughout the year, production could be more reliable and economical. Mary applied to the Smithsonian for a grant to learn more about the field of freezing and soon found herself working not just on fish but also on coral, working not just with marine scientists but with people who worked on cheetahs and lions, people who were concerned with preserving the genetic material of species whose futures were threatened.

By 2011, Mary worked out how to freeze and thaw coral sperm. She developed the technique for using those thawed sperm to fertilize newly released coral eggs, which successfully developed into functioning coral larvae. That same year, she established a Great Barrier Reef frozen repository, initially with sperm and stem cells from two species of *Acropora*. Today more than thirty species from twelve genera are in frozen banks, "an effective insurance policy to maintain the genetic diversity."

In 2017, along with scientists from Curaçao and Mote Marine Lab in Florida, Mary decided to try an unprecedented experiment in coral reproduction. Genetic work had shown noticeable differences in elkhorn coral near Curaçao and those a thousand or so miles northwest in Florida. These genetic differences indicated that elkhorns weren't reproductively connected with each other. Curaçao corals were breeding only with their neighbors. The same was happening in Florida. That kind of isolation is one way new species evolve. The question for the species was, Were elkhorn corals across the entire geographic range still able to produce babies? The answer wouldn't be found in the wild. In the sea, eggs and sperm from opposite sides of the Caribbean would never survive in the currents long enough to meet. Even if you captured the eggs and sperm, put them on a plane, and flew them across the Caribbean, the timing would be too tight. Unless they were frozen first.

Mary's team had already collected and preserved sperm from five elk-horn colonies in Tres Palmas Marine Reserve in Puerto Rico and from Elbow Reef in Key Largo, Florida. These frozen samples were sent to Curaçao. For two weeks surrounding the full moons in July and August, teams of divers boarded boats soon after sunset. After the sun dipped below the horizon, they flipped into the water and hovered near elkhorn corals. They watched and waited for the coral to spawn. Nearly two weeks after the August full moon, the nightly diving and waiting finally, literally, bore fruit. The divers placed small mesh tents over the animals. Each tent was topped in a tiny funnel that led to a small vial, in which the precious spawn was collected. On shore, the gametes were gently separated into vials of either eggs or sperm. Back in the lab, a massive in vitro fertiliza-tion got under way. Previously frozen sperm from Florida and Puerto Rico were mixed with eggs from Curaçao, as were fresh and previously frozen sperm from Curaçao. The entire lab was a test-tube orgy.

In the end, the most successful pairing was local Curaçao sperm mixed with Curaçao eggs. But some larvae survived from all of the different fertilization mixtures, proving that across the Caribbean, elkhorn coral are, for the moment, still able to reproduce. The test-tube-born larvae were then packed in water bottles, put into coolers, and shipped to Mote Marine Lab in the Florida Keys, where they would be coaxed into grow-ing into coral colonies. Before sealing them up, the Curaçao team slipped a little note written in colorful marker on graph paper into the package with the bottles. It read: SAFE TRAVELS! & GOOD LUCK TO THE CORAL BABIES!! MAKE US PROUD :)

That they did. At Mote, more than six hundred youngsters, gorgeous golden clumps of tentacle, survived. Mary called them "the largest wild-life population of cryopreserved animals in the world." In photos, the corals all look identical to one another. If you didn't know their extraordinary history—and weren't able to sequence their genes—you'd have no idea some of them were the reef version of designer corals. Those designer corals would have never been born in the wild, and for that reason the experiment itself has been called "assisted evolution." The ethics of such future efforts are still being debated: By mixing genes from Florida, where tougher environmental conditions have already been selected for tougher corals, with genes from Curaçao, where the impact from people has been less intense, might you be able to toughen up the population for what's sure to be a tougher future? Or conversely, because the gene pool in Florida is already low, could an inflow of new genes from Curaçao provide some genetic diversity for a dwindling population?

When I spoke to her, Mary said the designer corals were still living and growing in the aquaria in Florida. They looked healthy and happy, and it was past time for them to be returned to the sea. But scientists and restoration managers have been hesitant to put non-native coral on local reefs. "They're scared that they might be putting some monster out [on the reef]," Mary said. Given the continued decline in populations in the Caribbean, looming questions about whether and how to repopulate the reef with designer, genetically modified, or non-native corals need answers soon. The hesitancy reminded me of a conversation I'd had with Misha about the issue. "What is there to be afraid of? Coral invasion? That's what we want. That's what we want!" he'd said.

Freezing sperm for assisted evolution isn't the only aspect of cryopreservation that might be useful to coral restoration. Mary was also working on freezing embryos, dodging the trickiness of fertilization itself. After a dozen years, Mary was able to figure out how to freeze them, "but then I hit a boundary where I could not thaw them fast enough," she said. When ice crystals melt, they do so unevenly, throwing the concentrations of water inside and outside membranes out of whack. Too much water inside, and the cell can pop. Too much outside, and it can implode. But then in 2015, she met a colleague working on fish who had developed a new technology in laser warming. The secret was literally gold.

The particular wavelength of laser in the warming technique passed through biological material without damaging it or warming it up at all, which seems like it wouldn't be useful. But by first injecting multitudes of gold nanoparticles inside an embryo, scientists could use the metal to capture the laser's energy and then scatter heat evenly everywhere very quickly. The warming happened so fast and uniformly that there wasn't time for those problems with cells imploding or exploding to happen. When I asked what happened to the gold as the fish grew, Mary said it just remained inert. "They didn't have any adverse effects at all. I'm pretty sure the gold just sits there." Two years later, Mary perfected the technique on coral embryos.

Now she is hoping to take the gold-and-laser process a step further, working to freeze and then reanimate adult corals. This means freezing the full holobiont contingent: coral, algae, bacteria, and maybe fungi, too. "And that's a critical advantage," Mary said. "You're banking everything that makes a coral, a coral."

Because I've heard stories about human fertility, listening to Mary talk about frozen eggs and sperm, even if they were coral, was easy to understand. Even frozen embryos made sense. But adults? Isn't that what happened to Han Solo? Wasn't that the entire plot of Austin Powers? Wasn't that the stuff of science fiction?

As if intuiting my thoughts, Mary said, "Our process is extremely complex right now. But we will make it easier. We will make it faster, and we will make it restoration ready within the next couple of years." She paused, as if considering what might be lost otherwise, and added, "Definitely."

Meanwhile, the frozen bank where some coral genes already persist indefinitely keeps the door unlocked if we do decide to intentionally push the handle of evolution.

28

A Coral Named Romeo

Punta Cana, where Iberostar's coral lab was located, was home to elegant all-inclusive resorts. In contrast, FUNDEMAR, the lab that focused on coral reproduction, was headquartered an hour and a half away in Bayahibe, which was decidedly more casual. On the drive into town, we passed pastel-painted restaurants, and bars blasted music onto the streets, even though the sun was still high in the sky. A few hours earlier, Macarena, Iberostar's coral lab manager, had gotten word from FUNDEMAR that the coral had started to spawn two days earlier than predicted. I felt lucky to have arrived in time. Macarena, Megan, and I quickly readjusted our plans and made the drive southwest in hopes that we'd still be able to see some spawning. We parked around the corner from FUNDEMAR's yellow two-story office building. When we arrived at the front door, an insistent brown and black Chihuahua inspected us one by one.

Maria Villalpando, FUNDEMAR's coordinating scientist, emerged from a back room. She was slight, with dark, short hair and boxy glasses that framed her bright eyes. Someone later joked that she had the same energy as the Chihuahua: friendly and demanding. Maria greeted each of us with a kiss on one cheek. "It's been crazy," she said. "We are parents here, parents to millions and millions of coral babies." Looking down, she clicked her tongue at the dog. "It's okay, Sinojo," she said.

And that's when I noticed that the little dog had only one eye; *sin ojo* is Spanish for "without eye."

"He came to us sick," Maria said as explanation. "It was a joke at first, but it stuck."

I seemed to be in a place run by caring people with a sense of humor. Perhaps that was what it took to work on coral in the Caribbean.

Maria said that since the full moon five nights ago, their team had been up every night diving, capturing spawn, and fertilizing embryos in the lab. Two days after the full moon, *Dendrogyra cylindrus*, the pillar coral, spawned. Two nights later, it was the elkhorn coral, and last night the staghorn. All were endangered species. The lab was full of literally priceless life. Sinojo rolled over near Maria, who obliged with a belly rub. "Tonight, we think we will see three different species spawn." Maria stopped rubbing Sinojo's belly to scroll through her phone. "*Diploria labyrinthiformis* at 7:15, then *Acropora cervicornis* at 9, and then *Orbicella annularis* at 10." Three teams of volunteers and students would monitor the coral. "You are welcome to participate in any way you want," Maria said. But she wasn't going out tonight. She had earned a night's sleep.

At a dock around the corner, we climbed on a large dive boat with students and volunteers and motored a short distance to the coral nursery. The sun was starting to dip toward the horizon, and the bay took on a soft pink evening glow. Fishing boats bobbed on their moorings. In the distance, clouds gathered around the low island of Catalina. At the nursery, we tied off to a mooring, and the lead diver dropped a dive line and set out flashing lights to mark the surface. The first team of three suited up and jumped into the dusky water. The sun was now painting tangerine brushstrokes across the horizon.

My group, the staghorn team, would wait an hour before going in the water. Someone turned on music to pass the time and conversations popped up, most in quick Spanish, which I strained to follow. One of the students told an animated story that caught everyone's attention. He mimicked splashing and jerked his head back as if being hit in the face with cold water.

Everyone laughed, and Megan translated for me. He'd been watching a pillar coral, *Dendrogyra cylindrus*, and all of a sudden it blasted eggs in his face. "That's the way *Dendrogyra* spawns, in a cloud," she said from experience. "It feels mildly inappropriate."

The student said in English, "*Dendrogyra* is my favorite coral."

"Why?" I asked. I'm sometimes asked what my favorite animals are. I always have trouble settling on one, so I'm curious when someone else has a favorite.

"Because I like the way it looks," he said. "It always has its long tentacles out all day. It's the only one that does that." He pulled up a photo on his phone to show me a collection of pillars tinted a burnt orange. It was a stately, architectural coral. I could see why he liked it.

Another volunteer mentioned a particular *Dendrogyra* that had become famous for being a prolific spawner. "It even has a name: Romeo."

I laughed, admiring the romantic choice, and remembered that Maria had mentioned the species. "Did you see Romeo release sperm two nights ago?"

"Well," the second volunteer said, laughing, "Romeo is actually a girl. At least now she is."

"What?" I asked, confused.

"They can change sex," she said, as if it were the most common process in the world. I didn't know then that nearly all corals are hermaphrodites, meaning they are both female and male. It's not even clear if they have a sex chromosome: There aren't obvious chromosome pairs that look like XXs or XYs. For corals, gender fluidity is the norm.

Some species, like *Acropora*, release both eggs and sperm in tiny little packets. The waves break apart the packets, separating the eggs from the sperm, and then off they go to, hopefully, find a counterpart from another colony to fertilize. You might be wondering, What stops the egg and the sperm from fertilizing each other if they are released in packets together? Evolution has developed rather strict protocols to avoid self-fertilization (though it does happen). Eggs and sperm from the same individual are armed with protein-based protection that prevents them from forming an embryo. While these proteins haven't been studied in corals, results from urchins show that as few as ten changes in the hundreds of amino acids that make up the proteins lead to complete incompatibility. And evolution changes the proteins frequently. The compatibility proteins mutate three times faster than typical proteins.

In contrast, *Dendrogyra* like Romeo are sequential hermaphrodites. These animals start off as male and then turn female, or vice versa. This kind of transition is not unusual in the sea. In fish alone, sequential hermaphroditism has evolved more than half a dozen times. Clown fish (*Amphiprioninae*) like Disney's Nemo start off male and mature into female. (Nemo's dad in the movie really should have been a mother.) A small striped wrasse called a bluehead (*Thalassoma bifasciatum*) lives on coral reefs in the Pacific and switches the other way. It starts off life female living in a harem with one large male. If the male disappears, the largest female takes over that role, developing male organs and coloring. And then there are the Maori coral gobies (*Gobiodon histrio*). Just an inch and a half long, they are as photogenic as fish come, in bright teal with watermelon-pink dots and stripes. These monogamous couples have the ability to change either way. If the original pair is both male or both female, one switches so the pair is able to breed. The most gender-fluid animal I've heard of is the marine worm *Ophryotrocha puerilis*, a tiny creature with short, bristly legs first discovered in the Bay of Naples. When they are ready to mate, a larger female and a smaller male couple up. Except the worms don't grow at the same speed: The male grows faster than the female. So when he outsizes her, the pair changes sex, the new, larger worm taking over as female. That is, until the worm that was first-female-then-male becomes larger once more. At which time, they switch again. And so on.

But back to corals. They have even more gender-bending ways. In the coral species *Galaxea fascicularis*, the females make pink eggs that are capable of being fertilized. The male makes sperm, and also eggs. But his eggs can't be fertilized, which makes you ask, Why bother? And then there's this: "One time I saw a coral spawning," the student who had told me about Romeo said, "and in the same colony, one half [of] it was female and the other half male."

Leaning back against the dive boat's wooden bench, I looked past the edge of the boat's canopy and marveled at the sheer creativity and unexpectedness of nature. The moon hadn't yet risen, but the stars were glowing like hot coals in the night sky.

AFTER ABOUT AN HOUR, the first set of divers emerged from the darkness. They piled back on the boat speaking Spanish too fast for me to understand.

"Did anything happen?" I asked.

The lead diver spoke in English. "We were about ten minutes late to the party." They had seen only one coral colony spawn.

It was time for the staghorn team, which included me. I geared up, my fins, lights, and camera dangling like awkward ornaments. Megan was my dive buddy, and I followed her overboard. We kicked over to the lighted buoy, and she asked me if I was ready. I put my regulator in my mouth and signaled yes. I purged the air from my buoyancy vest and remained staunchly at the surface. I turned head-down, gave a couple of kicks, and popped butt-up right next to the buoy, like the guy in Mallett's painting *View of Coral-Divers*, though I had a wetsuit covering me. I willed myself downward. But gravity was having none of it. I hadn't brought enough weight with me. I felt a note of panic rise in my chest. Suddenly, it seemed very dark, the sea inky and forbidding. I was breathing too hard, huffing down the air in my tank, even as I remained above the water.

Megan resurfaced. "Are you okay?"

"I can't go down," I said, trying to hide my growing unease. "I think I need more weight."

Megan asked for more lead weight from the boat, which I added to my vest, flailing around on the surface of the sea as I did so.

Finally, I descended, though I was still breathing too hard in frustration and nerves. We followed the dive rope about thirty feet to a sandy clearing in the center of a coral nursery. I looked around, waving the beam of my flashlight. Arches made from wide mesh wire held thickets of staghorn coral. Staghorn coral also carpeted rebar structures that were similar to the Mars reef stars, but larger and set apart from one another rather than tied together in a web. Several wide rope ladders were held aloft with floats made from air-filled bleach bottles. Fragments of staghorn coral tied to strings dangled from their horizontal sections.

I trailed Megan around the nursery. She pointed to the coral and I looked, but I wasn't sure exactly for what. In the beam of my flashlight, small white bits flickered, tiny shrimp or other crustaceans. I squinted at the bits. Were some round and floaty? Were they egg and sperm packets? Maybe? No? Yes? Would I know? When you want to see something so badly, you tend to question all that's in front of you. Voiceless underwater, I realized I should have asked for more details topside.

For a little while, I was distracted by a flatfish, its slate-gray fins rippling in the sand. I explored a couple of bowl-shaped sponges and admired large waving sea fans. Again, I returned to the nursery area, peering at the coral—for what exactly, I wasn't sure. After less than an hour, I looked at my gauges. My earlier struggles had used a lot of air. I showed my gauge to Megan. She signaled that we'd need to head up soon. *Damn,* I thought. *I guess the coral aren't spawning after all.*

We surfaced and kicked back toward the boat.

"I guess we missed it," I said.

"No, they are probably spawning about now," Megan answered. "They were setting. They were very close."

"Setting?"

"Yes, the eggs were visible in the polyp's mouth. They were all set, and I expect about now they are spawning."

I was so mad at myself. Why had I screwed up my weights? Why had I let myself come to the surface when what I wanted to see so badly was happening down below? I splashed at the water. "Shit! I'm going to miss it."

"We should go back down," Megan said.

"How?"

"There are two extra tanks."

"Seriously? Can we use them?"

"If we go right back down, we won't have to worry about surface time. It'll count as one dive."

Quickly, we transferred our gear to new tanks and jumped back in the water. The sea had changed entirely. The water was alive with life. The coral had indeed begun to spawn. And while we missed the initial moments, we hadn't missed it all. I looked at a branch of coral and knew exactly what I was seeing. Tiny pinkish nearly round packages wandered away from the place of their birth. The motion was meandering, as if slightly uncertain they were ready for this moment. They were propelled by an internal buoyancy, the lightness of lipid sustenance the eventually fertilized embryos would rely on for this most important journey of their lives.

But threats were everywhere. The earlier activity of the little crustaceans in my beam of light had become manic, like a blizzard of hungry moths. The sea was soupy, thick with gelatinous life. Comb jellies, each with one or two pink egg packets already in their stomachs, wafted past my mask. Undulating chains of jellies called siphonophores, the size and width of a bracelet, wiggled by me. Each of the several dozen individuals in the chains held pink packets inside their clear bellies. A salp zoomed past. Inside its transparent body were many pink packets. Various sizes of fish picked and plucked at the bounty that surrounded us. All this fury made a racket. My ears filled with the clacking and snapping of indulgence and gluttony. A coral is not born into a safe or peaceful world.

As the staghorns' pink parcels disappeared from our flashlight beams to seek out their fate in the great sea beyond, the noise died down. But our dive had one more gift remaining. The boulder star coral, *Orbicella annularis*, was predicted to spawn soon. On the nearby reef, someone had placed a flashlight pointed at an *Orbicella* colony as big as a salad bowl. While we waited, we made a small circle around the area, peeking into

sponges and around nooks and crannies in the rubble. After a while we
circled back to that coral with a flashlight pointed at it. As if in a gasp, the
eggs started falling upward. They glinted off the flashlight beam so that
suddenly the coral really looked like it was encased in its own snow globe.
These eggs were larger than the staghorn's, and more buoyant and certain.
They lofted themselves upward in an insistent defiance of gravity. I looked
back at the mother colony. It was a species likely to fall prey to the SCTLD
epidemic. I tried to decide if what I had seen was hope, or futility.

JUST AFTER THE REEF FUTURES MEETING IN FLORIDA, I had taken Isy to
the psychiatrist for the first time. The office was downtown in a funky
complex tucked in among the live oaks. Inside, I filled out paperwork and
then flipped the light switch, which alerted the doctor that we were ready.
She appeared at the hallway door, long gray hair with sharp-cut bangs. I
shook her hand and called her by her first name. Her eyebrows raised at
the familiarity. I corrected myself, "Doctor." I hadn't made a good first
impression.

"I'd like to talk to Isy alone first," she said, ushering my daughter down
the hall.

I sat in the well-appointed waiting room, noticing how the bathroom
door blended in perfectly with the wood-paneled wall. After nearly forty
minutes, Isy returned. Her face was frozen, rocklike; my heart responded
in kind. I followed her down the hall to the doctor's office, and we went
over family history and medical history. As I recounted Isy's problems in
school, I saw tears brim in her eyes. The doctor prescribed an antianxiety
medicine. We should watch out for side effects: suicidal thoughts. The
medicine may or may not work. We wouldn't know for four to eight weeks,
and we had to ramp up slowly.

Back in the car, Isy's expression remained frozen. "Are you okay?" I
asked.

She melted. I hugged her as she sobbed. "It's so hard to have to talk
about all the things that are wrong with yourself for an hour."

"Did it hurt to hear things I said about what's going on with school?"

"Yes."

We were silent as I pulled out of the parking lot and onto the highway, silent as I neared the exit for her school. I couldn't bring myself to pull off. "Do you want to go for a walk instead?"

"Yes."

I drove to a trailhead, and we headed toward a clear spring that bubbles out of a limestone cliff—remnants of an ancient sea where forebears of today's coral once lived. As we rounded a turn, a beautiful swallowtail butterfly fluttered in front of us, black with yellow-orange dappled spots. We watched it loft above the trees and then stop mid-flight, fluttering awkwardly for a beat too long. I said, "Is it caught?"

"I think so," Isy answered, looking upward. "I think it's in a spiderweb."

And then I vaguely saw the strands of silk too. The butterfly bobbed in place, its wings extending through their full range of motion urgently. After a few moments, we both turned away. Neither of us wanted to watch its flapping stop. We trudged a few more steps down the rocky path toward the creek. And then unexpectedly, the black and orange butterfly swooped past us and above our heads, free.

"That butterfly is you, flying along," I said. "And you're snagged right now, and you have to fight. But you'll get free and fly onward."

"Stop it, Mom," she replied, walking ahead of me.

I knew my words were no balm. But I couldn't keep from saying them anyway.

29

Synchrony of Spawn

The Dominican Republic encompasses the eastern two thirds of the island of Hispaniola, which itself is the second-largest in the Caribbean. Its thousand miles of gorgeous coastline boast a diversity of ecosystems: wet meadows, swamps, estuaries, tidal plains, mangroves, springs, and coral reefs. Inland, you'll find the highest mountain, the lowest point, the largest freshwater lake, and the largest saltwater lake in the Caribbean. Beginning around 400 BCE, the Taíno, who trace their origins to the Arawak tribes of South America, farmed yucca, sweet potatoes, and maize, enough to support tens of thousands of people on the island. The Taíno made pottery, dyed cotton, and carved images on wood, stone, shell, and bone. They even developed rubber balls and built sport courts, which fascinated Columbus and his crew when they arrived in 1492. Of meeting the Taíno, Columbus wrote, "They love their neighbors as themselves, and they have the sweetest talk in the world, and are gentle and always laughing." Some of those sweet words may have even fallen from your own lips: *canoe, hammock, tobacco, barbecue,* and *hurricane* have roots in the Taíno language. But within fifty years of contact with Europeans, the Taíno population was reduced to mere hundreds. Disease killed the vast majority. Among the survivors, women were abducted and men were enslaved by conquistadors and put to work in gold mines. With no one working them, the fields that had kept people fed for generations produced no food. People starved. Many committed suicide; others fled.

Europeans began transporting enslaved Africans to Hispaniola in the early 1500s. Sugar plantations were introduced; then cattle ranches. French pirates, then Dutch raided the island. The next centuries were marked by power grabs between European nations interspersed with episodes of self-rule. The United States occupied the Dominican Republic side of the island, twice. Finally, following the assassination of authoritarian and murderous dictator Rafael Trujillo—with weapons supplied by the CIA—a constitution was signed in 1966.

Suddenly, the economy of the resource-rich country boomed, unemployment fell, and inflation held steady. That's not to say there still wasn't plenty of corruption and inequality, but conditions were ripe for the development of a tourism industry. That was just around the time that Ted Kheel, a noted New York civil rights activist and labor law negotiator, was looking for a new project. Thirty square miles of jungle and beach at the eastern tip of the Dominican Republic, a place known as Punta Borrachón—Drunkard's Point—or by the Taíno word Yauya, were up for sale. With a Dominican partner, Frank Rainieri, Kheel and other investors bought the land. They opened a rustic resort with ten two-room villas and a clubhouse. In 1978, Club Méditerranée (known as Club Med) bought a parcel of the thirty square miles to build a 350-room hotel, and it became clear that a real airport would be needed to serve the tourists who would come. So Kheel began what would be an eight-year process with three different governments to obtain authorization to build Punta Cana International Airport. In 1984, the first flight, a twin-turbo propeller plane from Puerto Rico, landed there. Today the airport is still owned by the company that Kheel and Rainieri started, known as Grupo Puntacana, and is the most successful privately owned airport in the world. My commercial flight landed there, and when I disembarked and turned a corner in the stairwell, bigger than life-sized photos of Ted Kheel and Frank Ranieri stared back at me. What they'd started had an enormous impact on the country. Six and a half million people now visit the Dominican Republic through that airport every year, more than visit any other island country in the Caribbean. Tourism accounted for 17 percent of the Dominican Republic's GDP in 2019, with spending by international tourists well over $7 billion. In recent years, the country's economy has been one of the fastest-growing in the Western Hemisphere.

Before traveling to the Dominican Republic, I spoke to Ted Kheel's great-nephew Jake Kheel. As the vice president of the philanthropic arm of the company, Fundación Grupo Puntacana, his job is to promote sustainability at the resort. But, he told me by phone, the seeds for sustainability were already planted by Ted Kheel and Frank Rainieri long before Jake showed up. They had used Taíno-inspired architecture for the airport that took advantage of breezes for cooling rather than energy-intensive air conditioners. They pioneered planting golf courses with a native Caribbean grass variety that vastly decreased water demand. Jake had developed a waste disposal system that diverted 60 percent of the solid waste from landfills and created compost for the golf courses and an organic garden used by the resort's many restaurants. Of most interest to me, when he realized that their coral reefs were critical to healthy beaches, Jake had spearheaded a project to build a vibrant coral nursery and employed local fishermen to work there as coral gardeners. While he wouldn't be in the Dominican Republic during my visit, Jake invited me to visit the resort's marine science laboratory where they had just completed a new "frag lab" for microfragmenting coral using David Vaughan's "Eureka moment" techniques. The day after the coral spawn in Bayahibe, Macarena and I drove over to see it.

Aquarist Noel Heinsohn met us when we arrived at Puntacana's coral lab and gave us a tour. Half a dozen long tanks were set up outside under a blue canopy. Noel pointed out the mother colonies that donated their tissue to be grown in small bits. To the side rested two saws used to slice coral into tiny pieces that grew at accelerated speeds. Each tank held hundreds of small nubbins of coral arranged on concrete plugs in plastic racks like pieces on a checkers table. "We're up to five species now," Noel said. "We hope to get up to a thousand fragments." A map on the wall showed the locations of restoration sites. He pointed to a spot not too far offshore. "That's the nursery where we went for the spawning last night."

"Did any of the coral that you've planted back on the reef spawn?" I wondered.

"Yeah. Yeah. That whole site is replanted coral. And yeah, they were spawning. Last night was just awesome."

Macarena said we'd been lucky to see a spawning as well in Bayahibe.

"I'm just curious," I said. "What time did they spawn?"

"We started our dive around nine fifteen, and I think we saw the [*Acropora*] *cervicornis* [the staghorns] about twenty-five minutes later. We surfaced around ten fifteen."

"Wait!" I said, trying to calculate what time it was after my bungled dive. "Isn't that when we saw it?"

Macarena nodded. "Yes, it was around nine forty."

"And we were, how far, fifty miles away? And around the corner of the island. And they spawned at the very same time?" I said, astonished.

"Yes, it's amazing," Macarena said, speaking for all of us.

Later that night, I walked out to the beach alone. I lay in the sand on my back and thought about how the corals hadn't conformed to our best predictions but had responded to a clock that they recognized effortlessly. They have no means that we can understand of communicating over distances: no phones, no televisions, no computers, of course. But also no eyes, no ears, no ability to even move. And yet, despite being separated by miles and miles of open water, the corals spawned in synchrony. I looked up at the sky. This night, the coal-bright stars from the night before were dimmed by the moon, which had just risen above the horizon, diminished from its full size and bathed in a pale orange. *How absolutely unbelievable*, I thought, *that a rock a quarter of a million miles away is the timekeeper for the coral's survival deep below the sea.*

I WAS RIGHT ABOUT THE MOON, mostly. But a lot of other things help set the coral's clock, too: the sun, the water temperature, wind, tides, nutrition, even rainfall. The most complicated and probable explanation is that corals measure the summer's warming seawater and integrate that with the period of darkness between the times the sun sets and the moon rises. Following a full moon, the period of darkness increases by about fifty minutes each day, and the corals respond to that change. How they accomplish all of that without a thermometer, watch, or light meter baffles me. But the fact that they do has put a crimp in coral science for decades. With just a couple of days each year to study spawning, fertilization, the development of embryos, the development of larvae, and larval settlement, making progress is slow. So aquarists and scientists have long tried to figure out how to convince coral to spawn more than once a year.

The successes have been inconsistent. It's one thing to look out at the world for cues that corals use to build their internal clock and then make predictions about when they'll spawn. It's quite another to reconstruct those celestial and oceanic gears indoors far from the sea. But beginning in 2012, Jamie Craggs, the aquarium curator at the Horniman Museum and Gardens in London, set out to do just that. The Horniman Museum was built by Victorian tea trader Frederick Horniman, who gifted his property to the people of London. While decidedly historic, Jamie pointed out during his acceptance speech for Aquarist of the Year at the 2019 Marine Aquarium Conference of North America, the Horniman isn't particularly spacious. He joked that he could sprint around the aquarium he oversees in just twelve seconds. It was fitting, I thought, that a clockmaker used time as a measure of space.

In 2015, Jamie shipped thirty-two *Acropora* from Singapore and the Great Barrier Reef to the Horniman Museum. If there were a reality show called *Extreme Homes: Aquaria*, the cribs he had built for these corals would be a featured episode. He constructed two systems, one for the Australian corals and one for the Singaporeans. Water was carefully controlled using chillers and heaters to reproduce the average annual temperature cycle for each location. Each aquarium was outfitted with an array of six different LEDs to simulate sunrise, midday, and sunset lighting conditions. The day/night light cycle and the daily-integrated intensity of light was calculated from NASA satellites to mimic the changing of the seasons at each latitude. A separate LED-controlled moonrise and moonset simulated the changing of the moon's phases. To make sure the intensity was perfect, Jamie cut Ping-Pong balls in half and fastened them over each light. It was still just a tad too bright, so he added three strips of tape. The corals were served a daily menu of amino acids, a baker's yeast solution, fish eggs, lobster eggs, and four different kinds of plankton. A support staff comprising two types of tangs, a rabbit fish, a butterfly fish, a wrasse, five hermit crabs, four urchins, and fifteen snails kept all variety of algae and pests at bay. And that was in addition to the pumps and filters, protein skimmers, baffles, and separate tanks full of special seaweeds that helped maintain the water chemistry as close as possible to the conditions on a reef in the middle of the tropical Pacific Ocean.

The corals arrived a month or two before they were due to spawn. Their gamete packets were already forming, and the corals released them as though they were still on the reef. This had been done in captivity before, so it was expected. But Jamie and his team learned something really important during that first spawning. Staying up all night was exhausting and unnecessary now that the corals were living in London. They reprogrammed the days and nights so that one species would spawn just after lunch and the others before teatime, or what we call dinnertime in the U.S. "Teatime is important," Jamie said. Now all that was needed was to keep the gears turning for another year.

In 2016, Horniman made marine husbandry history when all four species of corals spawned in London after a year in the aquaria. By 2019, the Horniman Museum had been able to spawn eighteen different species of *Acropora* corals.

Because of Horniman's cramped conditions, Jamie teamed up with the Florida Aquarium in Tampa, which recently built its twenty-two-acre Center for Conservation, largely dedicated to restoration and protection of Caribbean corals. By phone, Keri O'Neil, the project coordinator, told me that they currently have fifteen different species in their care, including one of the largest collections of refugees from SCTLD. She has managed to reproduce and grow so much staghorn and elkhorn coral that it's thought there is more genetic material of those species housed in one of her greenhouses than is left on the entire 380-mile-long Florida reef.

Working with Jamie, the Florida Aquarium had just completed installing four indoor coral spawning systems. The idea was to shift the start time in each tank by three months to produce one spawning per season. In the Dominican Republic, the same night that I wondered about the moon setting the clock for the coral, I checked my email just before bed. My Google alerts reported that the Florida Aquarium had spawned a new species of Atlantic coral in a laboratory setting for the first time. About thirty individuals spawned in synchrony, first the males and then the females about five minutes later. Keri O'Neil was quoted saying that while a couple of hundred eggs are collected from a typical spawn on the reef, the lab spawning produced thirty thousand. "It's really unheard of to have this many larvae," she told local news. "I'm really hopeful it means we'll have thousands of individuals to put back out onto the reef." I smiled when I read that the coral species that made these headlines was *Dendrogyra cylindrus*; Romeo's kin had made the news for spreading its seed.

ABOUT A YEAR LATER BACK IN AUSTIN, I got in the habit of joining Misha Matz's lab for a weekly meeting in which everyone gave a quick rundown on a recently published scientific paper. One of the postdoctoral fellows, J. P. Rippe, prefaced his presentation with: "I did my PhD work in Florida; it's my favorite place to dive. I always point out the *Dendrogyra* when I see them." Echoing the student on the boat in Bayahibe who had described why he liked *Dendrogyra*, J.P. said, "They are just amazing corals. They grow into huge pillars and they are the only coral in Florida that keep their polyps out all the time, so they look like fairy colonies." We were on Zoom, and J.P. clicked to share his computer screen. "A group of scientists from Nova Southeastern did a mapping project documenting each *Dendrogyra* colony in the Florida Keys. In 2014, there were eight hundred fifteen colonies," he said, scrolling to series of maps of the Florida Keys with locations of *Dendrogyra* marked by a colored dot indicating the coral's health—green for good, red indicating the presence of SCTLD, and black for dead. "I think this is the only map you need to see," J.P. said. In 2019, the map was dotted in black. Only fifty colonies still survived.

AS I CONTINUED TO FOLLOW KERI'S WORK at the Florida Aquarium, she announced success spawning another species, the ridged cactus coral (*Mycetophyllia lamarckiana*), in the lab. Then Jamie and his colleagues at the Horniman Museum published results that the baby *Acropora* that they had grown in their aquarium in 2016 had matured into adults over the past three years. And those mature *Acropora* themselves had gone on to spawn and produce larvae. This is known as an F2 generation. The coral's life cycle was, in the parlance of aquarium breeding, complete. But as the cycles of the moon and the sun and the forever shifting tides remind us, nothing is ever really complete. There are more species to understand, more coral to spawn, more larvae to grow and settle. Jamie told me that, unfortunately, none of those F2 larvae survived. So the lab experiments to refine the clockwork of reproduction will need to continue. And meanwhile, out on the reef, as the story of *Dendrogyra* so brutally shows, the stopwatch of disease and death just keeps ticking.

30

Coral Kindergarten

Although aquarists like Jamie Craggs and Keri O'Neil have done amazing work to boost the production of coral larvae, it is just one step on a very steep ladder. The odds are literally a million to one that any given coral larva will survive. Ecologists in the 1970s noticed that animals spanned a spectrum of parental strategies. On one end are big mammals. We have a long gestation period, put a lot of energy into our young, and keep them nearby for years. That means we can care for only a few in our lifetimes. On the other side are corals. They equip their eggs and sperm with the bare minimum and then toss them off, investing no further time or energy. Because each larva requires so little energy, they can make tens of thousands, hundreds of thousands, millions. But regardless of the strategy, only one or two of your offspring, maybe three or four in a strong generation, survive. Otherwise, the world would be overrun with a single species.

For coral, whose numbers are falling so precipitously, one way to boost populations is to boost the numbers of larvae. But it's not a sure thing. In one study in the Philippines, a million *Acropora* larvae were put out on a reef. After five weeks, it looked like the addition helped; scientists could see baby corals settled on tiles they'd put out to monitor the experiment. But a year later, the effect was gone. The coral babies didn't make it.

Another strategy would be to nurture larvae through settlement and let them grow big enough so that they can survive in the wild before putting them out on the reef. To see what that might look like, Macarena and I returned to Bayahibe, where we met FUNDEMAR's executive director, Rita Sellares. We climbed on her small boat and motored a little way out into the bay to an inflatable pool, about the size of a large dining room table, drifting on a mooring. It was covered with a blue tarp held up with PVC pipes, like a tent. About a dozen large laundry baskets zip-tied together rested in the bottom of the pool. Inside each basket were several dozen starlike ceramic forms about the size of my hand, covered in bumps and dimples designed to be welcoming nooks for coral larvae to crawl into. A few nights earlier, after a coral spawn and a massive IVF in the lab, the FUNDEMAR team had deposited hundreds of thousands of embryos in this floating incubator. Compared to the cost of building and maintaining indoor aquaria, the ocean water surrounding the inflatable pool controlled the temperature while the sun took care of the lighting, all for free. This entire apparatus—the inflatable, the tenting, and the tetrapods— were an experiment in scaling up coral restoration. The system came to FUNDEMAR through a partnership with SECORE International, another group whose name is a concatenation, SExual COral REproduction, and which had stood out to me at the Reef Futures meeting.

Dirk Petersen, whom I spoke to by phone before my trip to the Dominican Republic, founded SECORE. In 2002, while working on his PhD at the Rotterdam Zoo, Dirk developed techniques for teasing coral larvae into settling down on bits of rock so that they could grow into small juveniles. The idea wasn't new for marine scientists—they had studied larvae and how they behaved for decades—but, he said, "to make the bridge to captive breeding in the public aquarium world, this was a big new thing. Aquarists were seeking a new challenge, and that was applying sexual reproduction [of coral] to aquaria."

Dirk, along with Bart Shepherd, the director of the Steinhart Aquarium at the California Academy of Sciences, recognized that one bottleneck to using aquaria-bred coral larvae in restoration was the one that Jamie Craggs was trying to solve by shifting the coral's timing gears to spawn throughout the year. But a second, and just as narrow, bottleneck was planting coral back on the reef. To restore one square kilometer of reef, you'd need four million coral colonies. "What's realistic?" Dirk asked. "If ten people work for eight hours a day, you'd end up with several hundred years to plant four million corals. We need very smart solutions for cultivating coral and getting them established on the reef."

A potential part of that smart solution was the tented floating pool moored off Bayahibe. SECORE called them "floating coral kindergartens," because they were safe spaces where coral embryos could grow and mature. The other part of the solution was sitting in the bottom of the laundry baskets, those strange-shaped tetrapods that looked like toys you'd find in a kindergarten classroom. One of the FUNDEMAR scientists pulled one from the raft and placed it in a plastic tub so I could look at it more closely. I put on a pair of glasses that blocked all but green light, and shined a blue light onto the tetrapod. I could just make out very, very tiny green dabs. That was fluorescence coming from the algae in the coral, evidence that the larvae had begun to build their their homes.

Later, when we visited the shipping container that Rita, Maria, and the rest of the FUNDEMAR team had converted into a wet lab, I would have a better view. I knelt on the metal floor and peered at eye level through the glass of an aquarium. Sinojo, the one-eyed Chihuahua, wandered over to inspect me, and I reached out to pet him distractedly. My eyes focused down to the size of a snowflake, and I saw a thin blizzard of very tiny but determined coral larvae. Shaped like fat rice grains, with one end a bit larger than the other, they didn't fall but swam constantly, with the rounder end propelling them in a kind of corkscrewing motion. Most were headed downward, but without sinking. It looked like it took effort to descend, and I remembered how I'd felt the night of the spawning with too little weight. Except the larvae weren't struggling or flailing; there was a lazy grace to their motion. I watched mesmerized as one larva made a choice, perhaps the most important choice it would ever make in its life. It picked a spot to settle. In her book *Lab Girl*, Hope Jahren describes the moment for a plant's seed, which similarly has just one chance to get everything right:

No risk is more terrifying than that taken by the first root. A lucky root will eventually find water, but its first job is to anchor—to anchor an embryo and forever end its mobile phase, however passive that mobility was. Once the first root is extended, the plant will never again enjoy any hope (however feeble) of relocating to a place less cold, less dry, less dangerous. Indeed, it will face frost, drought, and greedy jaws without any possibility of flight. The tiny rootlet has only once chance to guess what the future years, decades—even centuries—will bring to the patch of soil where it sits. It assesses the light and humidity of the moment, refers to its programming, and quite literally takes the plunge.

So, too, a coral larva has one existential job: to settle, ending forever—years, decades, centuries if it is lucky—its ability to seek a better spot. It won't experience frost or drought, but it will endure heating and cooling, lightness and darkness, the scratching teeth of snails and urchins, the biting jaws of fish, the boring of worms and mollusks, and never again will it be able to flee.

Scientists do not know exactly what convinces a coral larva to plant itself. But they have ideas. A nook is good, especially if it's tilted a bit. A larva likes a bit of sun, but not too much. It's a balance between limiting exposure by hiding in the shadows before it builds a protective skeleton and ensuring enough sunlight to power the algae. Larvae seem to be encouraged by a kind of algae called, appropriately, coralline algae. It is usually pinkish or white and forms lichen-like crusts on rocky surfaces. Certain types of bacteria also seem to encourage settlement. Perhaps the presence of these fellow surface dwellers signals decent lighting and not too much scraping by predators. There's even evidence that the coral larvae listen for the sounds of the fish and crustaceans, the music of a bustling reef.

Staring into the floating pool full of larvae, Rita said that in a few more days, when the corals reached a few millimeters in size, the tetrapods would be placed onto the reef. SECORE had engineered the pointy

shapes to catch on crevasses and divots and lodge there. In just an afternoon, hundreds of thousands of tiny corals would be planted. Perhaps it
was the kind of scale-up that could have an impact. A study from the first
coral kindergarten experiment in Curaçao found that the tetrapods decreased planting time by as much as sixty-six-fold and costs by as much
as eighteen-fold. After a year, about 10 percent of the corals were still alive
on the tetrapods, and two thirds of the tetrapods still contained at least
one coral colony. I thought back to the feeding frenzy of the night before.
Protecting the corals until they were settled down bypassed some of the
most dangerous parts of its life. Just like we try to do with our own children, the hope was to give them the best we could before we launched
them into the world.

MY LAST EVENING in the Dominican Republic, I made a dinner reserva-
tion at one of the upscale restaurants of the Iberostar resort. After my
deconstructed pumpkin soup was expertly poured into my bowl, my
phone rang. It was a FaceTime call from home. I answered in a whisper.

"Hi, Mom!" Isy said, her round face filling my screen.

"Hey, Sweets! I'm at a fancy restaurant and not sure I should be on the
phone," I said, glancing around at the other diners, who actually didn't
seem bothered at all.

"Okay, I'll go. Call me later?"

"Sure. Want to see my soup?" I trained the phone on the picture-
perfect dish.

"Not really. Bye. Love you."

"Bye. Love you."

As I savored the sweet and spicy flavors, I thought of Isy. Her blood-
work still showed very elevated antibodies to strep, even after rounds of
antibiotics. And she had been retreating into a world of ever stricter and
more mysterious rules. "Contamination" lurked everywhere.

I had recently thrown a small party for some other writers at our
house. Keith was out of town, so I asked Ben to pick up Isy from dance.

On the way home, he was hungry, so he stopped at a burger joint's drive-through. He hadn't realized that he'd gotten too close to the school where things had gone so wrong, a place that was now "contaminated."

My phone went crazy with texts:

> We are at the toxic school! I can't handle it!
>
> I am going to cry! I need you now!
>
> Mom! Help!
>
> Are you having a party?
>
> I can't I can't!

When they got home, I slipped away from the wine and the chatter and met Isy in my bedroom. She collapsed in my arms, sobbing. I held her, but also wondered how long I could hold her, with a house full of laughing guests.

"You've passed that spot a million times before and been fine," I tried, knowing logic had no power over her thoughts.

"I'm contaminated. I'm so contaminated. I need to shower. I need to get clean," she cried into my arms.

"The school was closed. No one was even there," I offered.

"I hate this. I hate this!"

"Go ahead and take a shower," I said, even as I knew I was only strengthening her response to the trigger. I slowly walked downstairs and rejoined the party, the sound of laughter mixed sourly with that of water running from Isy's bathroom upstairs. It was the sound of my failure, my inability to protect her from the agony of her thoughts.

31

Category 5

I left the Dominican Republic on August 25, 2019. Two days later, the weather service predicted that tropical storm Dorian, which was plodding its way across the Atlantic Ocean toward the Caribbean, would make landfall, likely as a hurricane, possibly as a big one. It was aimed directly at the eastern end of Hispaniola Island, exactly where I'd been. Sitting out in the water, I knew, were the floating SECORE pools full of baby coral that FUNDEMAR had so carefully nurtured over weeks of sleepless nights. They needed another week at least to settle safely and build their tiny skeletons. The pools would never survive a major storm moored as they were in the open sea.

I wrote to Maria for updates. She said that they had been forced to move the larvae from the pools into the shipping container lab to ride out the storm. I signed my emails, "Wishing that the storm turns." But when it comes to the sea, I should have been more careful.

The day Dorian made landfall on the island of St. Lucia, it did turn, veering northward away from both the Dominican Republic and Puerto Rico, which had been devastated a year earlier by Hurricanes Maria and Irma. The storm moved past the Virgin Islands as a Category 1, and then over the warm Caribbean waters, which fueled its fury, accelerating it to Category 5 status. When Dorian collided with Great Abaco in the Bahamas, the sustained windspeed was 185 miles per hour, tying the record for the strongest hurricane to make landfall, a record that had held since 1935. New data suggests that there are more tropical storms than there used to be. Of the top eight seasons with the most named storms, all but two were in the past two decades. And what has been certain for some time is that heat is the fuel that spins larger, more intense, and more damaging hurricanes. The year 2020 was the third-hottest for ocean temperatures on record to that point, followed by 2016 and 2019.

Dorian was a horrific beast of a storm. It sat on the Bahamas for two days, moving at a grueling single mile per hour. When it finally lumbered northward after that incessant pummeling, upward of seventy thousand people were left homeless. It macerated houses, trees, stores, marinas, hotels, yachts, schools, hospitals, power plants, power lines, water facilities, airports, factories, farms, and anything else that got in its way, including Coral Vita.

As the storm sat on Grand Bahama Island hour after hour, I scanned through Coral Vita's Twitter feed. On September 1, they posted a 17-second clip of the farm, rows of sea tables inundated with rain, filtration systems and offices still standing, overlooking a gray sea as palm fronds swept fitfully in the growing wind. They wrote: "#Dorian . . . made it to the #coralfarm in #GrandBahama. Shutting it down and heading to the home bunker. 🙏 for the people of the #Bahamas, and get supplies and help ready to send in."

By September 3, Coral Vita had turned themselves into a triage unit. They'd set up a relief fund on GoFundMe and were working to get supplies where they needed to go. They said they had loaded a truck with chain saws and wanted as much emergency aid on hand as they could muster. They connected the storm's devastation and recovery to the need for resilient coastal ecosystems. The plea worked. People cared. Over $50,000 was donated in one week. In the following days, the Coral Vita team rescued more than a dozen people. They were part of the first group to make their way to remote villages, bringing food and water to places that had been completely cut off. The official death toll from the storm remains uncertain, but hundreds of people were either killed or unaccounted for after a month. Infrastructure damage was estimated to be more than $7 billion.

By September 20, the situation on the ground shifted. Coral Vita handed off aid efforts to the professional groups that finally arrived. Sam and Gator turned their attention back to corals. Even though Coral Vita had built its farm with storms in mind, locating the aquaria's foundations four feet above what was predicted to be the highest high tide of the century, Dorian's surge was thirteen feet higher. Blue-painted cinder blocks used to support tanks were strewn about like discarded toys across the floor. PVC pipes that ran the filtration system stuck vertically out of the ground connected to nothing. Windows were blown out of the lab; a metal door was ripped near the handle like it was a piece of paper. Every tank was washed away along with the entire coral stock. One aquarium, a fiberglass basin eight by four feet and painted bright blue, was found in a village thirty-five miles to the east. Word came that another tank had washed up in a nearby backyard. A friend with a helicopter spotted a few tanks amid overturned cars and the debris from houses. By late October, fifteen tanks had been located.

For two years before the storm hit, scientists from the Perry Institute for Marine Science based in Vermont had been working to survey four hundred coral reefs around the Bahamas, by both in-water diving and high-resolution photography. Two months after the storm, they were in the water again, assessing the damage with the rare perspective of a very clear "before" picture in hand. When storms like Dorian stir the water, corals can be turned over or smashed. The fragments roll around, preventing the reef from reestablishing itself. Branching corals like elkhorn and staghorn break easily, but they also grow back quickly. Slower-growing corals like brain corals and star corals have more sluggish recoveries. Following stress, corals might bleach. Stirred-up sediment can bury corals alive. The Perry Institute scientists expected a mirror image of the devastation that they had seen on land, but that's not what they found. Only about a quarter of the Bahamian reefs suffered severe damage.

Some places might have even benefited because the storm tore away sea-weed that competes with corals for space. Because healthy reefs remained, recovery could be accelerated with coral restoration and protection. "There is hope for recovery of most coral reefs in the area," they wrote.

In an email following the storm, an exhausted-sounding Gator explained why Coral Vita would rebuild. "Over and over these brave Bahamians pointed to the land around their homes, often saying that the mangroves that grow there are the only reason they're still alive. Similarly, it is no coincidence that the southern half of Grand Bahama remains relatively intact, with the southern shore rung by a barrier reef about a half mile out to sea."

PART VII

WASHINGTON, D.C.

32

X-tinguished

After the Reef Futures meeting, I had joined the online coral XPRIZE community and avidly followed their forums. There was much lively debate, largely about how to frame the rules. But about a year later, just around the time I expected the XPRIZE to go live, the community dialogue went quiet, and then it disappeared entirely. I sent a note to Matt Mulrennan, who had introduced the XPRIZE at the meeting, asking for an update. As I had begun to suspect, the problem was funding. Originally, the country of Abu Dhabi, which has a sovereign wealth fund, was going to support the XPRIZE. But that effort hadn't worked out. Other funders hadn't been identified.

This is a chronic problem. Ever since the possibility of launching humans into space was broached, humanity collectively turned its attention away from the dark, unknown seas in favor of the vast, mysterious heavens. Celebrated author and ocean enthusiast John Steinbeck recognized the shift back in 1966 in the midst of the race to put a person on the moon when he penned an open letter to the editor of *Popular Science* with a plea to create a NASA for the oceans: "I know enough about the sea to know how pitifully little we know about it. We have not, as a nation and a world, been alert to the absolute necessity of going back to the sea for our survival. I do not think $21 billion, or a hundred of the same, is too high a price for a round-trip ticket to the moon. But it does seem unrealistic, unreasonable, romantic, and very human that we indulge in these passionate pyrotechnics when, under the seas, three-fifths of our own world and over three-fifths of our world's treasure is unknown, undiscovered, and unclaimed."

Despite elegant prose like Steinbeck's, funding for the oceans has always paled in comparison to that of space. In the United States, NASA's 2021 annual budget was over $23 billon, fivefold more than the National Oceanic and Atmospheric Administration's at $4.6 billion. Illustrative of this long-term funding gap is the fact that in 2013 the best maps of our own seafloor had a resolution of three miles. The best maps of Mars had a resolution of 330 feet. Internationally, the situation is no better. In 2015, the United Nations developed a list of seventeen Sustainable Development Goals that we as humanity should strive to achieve by 2030. They include noble, important, and difficult aspirations such as eliminating poverty and hunger, providing quality education, guaranteeing clean water and sanitation, and ensuring decent work for all. Goal number 14 is "Life Below Water," defined as "Conserve and sustainably use the oceans, seas and marine resources for sustainable development." Of all the United Nations Sustainable Development Goals, "Life Below Water" garners the least philanthropic funding of all, just 0.56 percent.

But in this very discrepancy between sea and space funding, perhaps there was an opportunity for the coral, Matt said. XPRIZE was courting donors with an interest in space exploration. "We talk a lot about needing to save Spaceship Earth. It has a life-support system already installed, but the warning lights are going off and telling us we need to make repairs. There's a realization in the space community that we have a crisis here on Earth. Dire headlines are stacking up: reports of extreme weather, mobilization of climate activists in the streets, and a lack of leadership at the government level. We can't count on them to step in. The IPCC [Intergovernmental Panel on Climate Change] report released in November that predicted 90 percent loss in coral, even if we hit 1.5 degrees Celsius [warming above pre-industrial levels], that really hit people in the face."

WITH THE XPRIZE ADRIFT, I decided to check in with some of the potential contenders that I'd met at the Reef Futures meeting a year earlier. The
Mars effort was unquestionably inspiring. It was close to the pinnacle of
what restoration could accomplish in the prime location. But its rollout to
other locations had slowed. One more restoration had been installed in
Mexico, but Alicia told me that permissions to install on the Great Barrier
Reef had been slow to come. The government was concerned about the
use of plastic zip ties on the reef and had opened up a period of public
comment. In the meantime, Mars was researching biodegradable alterna-
tives to zip ties. Everyone was just waiting.

While spots of promise existed in the Caribbean—IVF projects
like FUNDEMAR's and SECORE's continued to make coral babies, and
nursery-based groups like Mote and the Coral Restoration Foundation
continued to grow and replant coral on the reef—environmental conditions were worse than a year before. SCTLD was still raging in Florida
and spreading well beyond. Iberostar's Megan Morikawa had emailed
that the disease had been detected on the nursery reefs where I'd dived in
the Dominican Republic, making her work preserving genetic diversity
more critical. Mote Marine Lab, which had advanced the ideas of fragging
and reskinning, had also shifted their attention to saving healthy coral
from the onslaught of the disease, along with many of the aquaria and
universities throughout Florida. Coral Vita in the Bahamas was still in a
physical shambles, rebuilding from the devastation of Hurricane Dorian.

It was time for a dose of optimism. I decided to check in with Richard
Vevers.

RICHARD RETURNED MY CALL while I was walking through the hardware
store looking for a replacement garden-hose faucet. So I sat down on a pile
of boxes in the plumbing aisle to talk about coral. It wasn't the first time
doing research on ocean issues from central Texas has felt disorienting.

The Glowing Glowing Gone campaign, which used the color palette of fluorescent and dying coral, was gaining traction, he said. The World Surf League's Tahiti Pro Teahupo'o had been completely taken over in the colors of glowing corals through a partnership with the athletic clothing company Hurley. One of the leading international dive-training companies, PADI (Professional Association of Diving Instructors), had signed on and was using the colors of coral in their marketing. Along with the vice president of Pantone, Richard had recently presented the coral colors to "thousands of screaming creatives" at the Adobe MAX conference in Los Angeles. But he'd also heard from a number of companies that he'd pitched the campaign to that coral already seemed like a lost cause. "We're considering shifting our focus to highlight climate change," he said of The Ocean Agency's future direction.

Just a month earlier, the world had been captivated by Greta Thunberg, a Swedish sixteen-year-old who had spoken with passion and anger before the United Nations, accusing the adults in the room of destroying her childhood and stealing her future. She said they had denied her a habitable planet in order to chase "the fairy tale of ever-expanding economic growth." Greta galvanized activists and especially youth. After the UN speech, climate strikes took place in hundreds of cities around the world. Hundreds of thousands of people took to the streets to demand action to limit carbon emissions. I understood what Richard was seeing. Rather than work on coral as a gateway to action on climate change, his efforts might be better spent working on climate change itself, the existential and underlying threat to coral.

Richard said, "The half a million who show up at marches are great, but what needs to happen is that we need to figure out a way to unlock the voice of the two hundred million people in the U.S. who already believe that we need action on climate change."

THE LAST PERSON I CIRCLED BACK TO was the man who had kicked off the Reef Futures meeting with so much positivity, Tom Moore. He had recently had something of his own breakthrough. He recognized that the Florida coral restoration community had made an error talking about the coral reef crisis. In the past, when people asked him, "How much money do we need for restoration?" the answer had always been "A lot." Tom said, "We were afraid to define what we needed."

So, along with a team of experts, Tom had developed a well-defined road map to restore seven historically and environmentally important reefs, called Mission: Iconic Reefs. "We are going to take places that are currently the poster child of decline and make them iconic again," Tom said. The vision was for these seven reefs to be restored to conditions of fifty years ago. They would act as habitats for fish, crustaceans, mollusks, echinoderms, worms, and other reef denizens. The project would roll out in two phases over twenty years, inspiring divers and thus becoming an economic driver. The first phase started with three years of fundraising, and then seven of active restoration. Critically, he was able to say what the price tag was: $100 million for the first phase and $150 million for the second. "We are looking for public and private funding," Tom said. "This is an opportunity for private money to contribute to restoration—the people who live in these communities and the businesses that depend on these reefs that could invest."

"It sounds kind of like a capital campaign for the reef," I said, thinking of the efforts I'd heard of where nonprofits like schools and churches raise large sums of money for renovating buildings.

Tom said it was. "The restoration documents look a lot like construction drawings. We are going to define the problem, define the solution, and give people a choice if they want to make it happen. An important thing to recognize is that historically there was a huge debate about coral restoration. Does restoration have a role in coral research? Over the last year, we've shifted the narrative. Restoration has a role. If we don't stabilize the system, there won't be a system when we finally get around to being able to do something about it."

I asked the question that Richard had posed: Should we be focusing on climate change instead of coral?

"One thing that gets lost often when we talk about restoration—and it's a real downfall to the message of restoration—is the idea that we have to address climate change and ignore coral. People are capable of understanding the nuance. I'm convinced that we are never going to be in a situation where corals drive us to address climate change. What's ultimately going to get us to address it is flooding in New York and Florida, massive changes to our economic reality. But losing corals is not an option." And he repeated the line I'd heard so many times during my conversations with coral people. "And giving up is not a solution."

AROUND THE SAME TIME, Florida's Coral Restoration Foundation, which had pioneered the coral orchards, took over the giant screen in New York's Times Square. The video opened with an image of a brain coral. It referenced a famous public service announcement from the '80s about drug use: "This is your brain," a man said, holding an egg. Pointing to a hot skillet, he said, "This is drugs." He cracked the egg against the skillet and dropped it onto hot oil. As the egg started sizzling, the man added, "This is your brain on drugs." After a poignant pause, he asked, "Any questions?"

In Times Square, the Coral Restoration Foundation video zoomed in on the brain coral's stark whiteness. The text read, "This is a brain coral." After a pause, the text flashed, "That's been bleached." Then it read, "Due to global warming." After a moment, the text asked, "Any questions?" The image switched to a coral orchard made of PVC pipe trees. Fragments swayed from the branches. "We have answers," the text read, and the Coral Restoration Foundation's website address was given.

But the answers seemed vague to me. The coalescing of energy that had dominated the Reef Futures meeting had become diffuse over the previous year. Rather than act as a way to drive coral restoration efforts forward, the XPRIZE had evaporated. In that vacuum, Richard and Tom represented the opposite sides of a spectrum of responses. Richard was turning away from his focus on coral to the larger global issue of climate change, and Tom was becoming hyperfocused on just a few reefs where he might have meaningful impact. But both confronted the same problem: How would we pay for it?

33

The Lifeboats

When Richard Vevers and I were in Indonesia, he told me that, as he'd traveled the world's reefs, he'd realized that, given the limited funds for coral conservation and restoration, it would be impossible to protect them all. After scribbling out numbers on the back of an airline napkin that added up to trillions of dollars, I was largely in agreement. Were there certain reefs, he wondered, that made more sense for investment? He reached out to some of the world's most prominent coral researchers to help answer the question. The result was a project called 50 Reefs, which aimed to identify the planet's fifty most critical reefs.

The first step was to assemble a dataset of coral reef metrics. The team identified thirty such characteristics, including historical seawater temperatures, predicted future seawater temperatures, storm frequency, and the abundance of coral larvae, among others. To analyze the dataset, they looked to the world of finance, borrowing a technique called modern portfolio theory developed by economist Harry Markowitz, who won a Nobel Prize for the idea in 1990. The theory is widely used by stock market portfolio managers who try to pick a range of stocks that maximize returns, minimize risk, and diversify investments. Similarly, the 50 Reefs project's goal was to pick reefs that maximize health, minimize risk, and diversify locations around the world. Reefs that rose to the top included places in Indonesia, regions of the Great Barrier Reef, Tahiti, and the northern Red Sea. Notably, reefs important to many people were left off the list. The entire Florida Reef was not included, nor was any reef in Hawai`i, Micronesia, or the Mesoamerican Reef of eastern Mexico and Belize. Such omissions raised ethical questions. Picking and choosing implies leaving out: What about the fifty-first reef?

The philosophical conundrum posed by the 50 Reefs project isn't new. In 1974, Garrett Hardin—that same thinker who coined the Tragedy of the Commons—published an essay titled "Living on a Lifeboat," which frames "lifeboat ethics" like this: Suppose you are in a lifeboat that contains fifty people, although the capacity is sixty. For safety, you are advised to keep ten seats free. Yet a hundred people are swimming in the water all around you. What do you do? Hardin says you have three choices. First, you could help everyone. But the lifeboat will sink and everyone will drown. That's a bad choice. Second, you could decide to choose ten others to save. But how do you choose? That's the 50 Reefs scenario. Third, you could refuse to let anyone else aboard to ensure your boat stays afloat, which Hardin believes is the best answer. It goes without saying that it's also the most cold-blooded and abhorrent.

Hardin recognized that humans don't decide to refuse other people a place in the lifeboat without guilt. If you feel guilt, "get out and yield your place to others," he wrote in that chilly manner of his. "Such a selfless action might satisfy the conscience of those who are addicted to guilt, but it would not change the ethics of the lifeboat. . . . The net result of conscience-stricken people relinquishing their unjustly held positions is the elimination of their kind of conscience from the lifeboat. The lifeboat, as it were, purifies itself of guilt." Reading that word "purifies" makes my toes curl. It calls to mind eugenics, which Hardin later endorsed. The questions Hardin forgets to answer are these: Isn't guilt a piece of what makes us aware of injustice, and isn't striving for a better world what makes us human? Isn't living in a lifeboat after you've watched those around you drown unbearable?

Given these ethical problems, it's not surprising that the 50 Reefs study drew criticism. In addition to the fifty-first reef problem, objectors said that the work left out important factors like genetic diversity and local social and economic support for the reefs. And, skeptics said, an analysis like that of 50 Reefs was an excuse to dodge the underlying problem of climate change. The scientists involved in the 50 Reefs project acknowledged the shortcomings, responding that the goal was to act as a starting point, one to be used in concert with the reduction of carbon emissions.

Even with the criticism it drew, the 50 Reefs project brought a kind of focus to coral reef science that had been missing, and resulted in one of the largest donations to support coral reefs ever made. Bloomberg Philanthropies, established by businessman and politician Michael Bloomberg, committed $86 million to the Vibrant Oceans Initiative, funding coral reef conservation in ten countries, including Australia, the Bahamas, Indonesia, Tanzania, and the United States. The work mostly focused on tracking illegal fishing and enforcing already existing marine policies. Around the same time, Vulcan Inc., established by Microsoft cofounder Paul G. Allen, who was also an early investor in SpaceX, committed between $3 million and $4 million per year to develop a high-resolution atlas of the world's shallow coral reefs. The project is called the Allen Coral Atlas.

A few months after Lizzie Mcleod—The Nature Conservancy's global coral reef lead—and I had spent the morning eating breakfast tacos and fighting a microwave fire, she moved from Austin to Alexandria, Virginia, to be closer to the Washington, D.C., headquarters where she worked. We'd stayed in touch, and in February 2020, Lizzie invited me to visit. I questioned whether I should go, for two reasons. Isy had become even more isolated from the world. In the past two months, she'd stopped going to dance, which had been one of her few remaining reasons to leave the house. In the past month, she'd also stopped going to the alternative school where she'd been hanging on by a thread. She was shutting down. We were considering more aggressive medical treatments. And she had asked us to look into more intensive therapy at a residential mental health program that specialized in adolescent OCD. Indicative of the lack of options for young people suffering from mental illness, only two such programs existed in the country, with a total of thirty-six beds. Isy had just been wait-listed at one of them in Wisconsin, and I worried about leaving town while we were trying to make major decisions about her mental health, including how much they would cost. As with the health of reefs, providing for mental health is too often absent from our plans. Just as concerning, alarms about the spread of COVID-19 were beginning to sound. A few cases had shown up in Washington state, and The Nature Conservancy had just made going to the office optional. Even weighing the negatives, I convinced myself it made sense to travel to meet experts face-to-face. Or maybe I was looking for a reason to escape the looming threats.

I met Lizzie in an arty, airy restaurant just outside of Washington, D.C., that served endless cups of coffee in heavy white mugs. We were joined by Helen Fox, who headed up the field operations team of the Allen Coral Atlas, which was based at the National Geographic Society. Helen's graduate work in Komodo National Park in Indonesia was one of the first studies to look at large-scale coral restorations from blast fishing, and one of the few that included long-term monitoring. As the three of us sat down together and began to talk, our similarities emerged. We were close in age, all with doctorates in marine science. Our work had given us the gift of seeing the world but had also involved disappointments and compromises in our personal lives. We had all found our way into the kind of motherhood that meant juggling our love for our work with the demands of being a parent. I felt a camaraderie as we talked about our struggles in science as well as a shared recognition of our grit and determination born of those struggles. In the midst of these women, I also felt an alliance of optimism, buffered by caution, for the future of the seas.

Like the reefs themselves, Helen said, the Allen Coral Atlas was a complicated collaboration among partners, coordinated and led by Vulcan Inc. High-resolution satellite imagery was captured by a company called Planet. Planet was founded by three NASA scientists who launched a fleet of hundreds of shoebox-sized satellites that circle the globe every day. They sent the imagery to a group of scientists at Arizona State University, who cleaned and stitched it together. This imagery went to another group at the University of Queensland in Australia, who turned the images into maps of coral reefs and nearby habitats. National Geographic then reached out to managers and decision makers on how best to use the information in the maps.

"How far along are you in the project?" I asked.

"Three fourths of the world's coral reefs still haven't been mapped in detail," she answered, which seemed like a lot to me. Maybe it shouldn't have, given what I knew about ocean funding.

Stirring sugar into my second cup of coffee, I asked how Helen envisioned people using the Atlas once it was complete.

"In addition to helping with planning the boundaries of marine protected areas, I see the Coral Atlas acting as an auditing function and for testing hypotheses," she said. One kind of hypothesis could be searching for reefs adjacent to deep pools of cooler water or other unique local characteristics that might provide coral a refuge in warming seas. Another sort of hypothesis could be looking for reefs experiencing specific problems. The collaborators at Arizona State University were already adding in "brightening alerts and turbidity alerts, because those are two things that will show up in the satellite imagery." Brightening would signal bleaching, and turbidity would signal pollution. She said that more detailed mapping could propel the field beyond the first-stab analysis performed by 50 Reefs, pinpointing reefs that are surviving better in future ocean conditions.

Since she'd broached the subject, I asked, "What do you think of 50 Reefs?"

"It was a great first step," Lizzie said.

"But it had limitations . . ." Helen said in a cautionary voice.

"Like, genetic diversity . . ." Lizzie said.

". . . and reefs that were left off . . ." I said.

". . . and thermal microhabitats could play a big role . . ."

We were all talking over one another, our thoughts about reefs tumbling about like waves.

Lizzie slowed the conversation down. "It is important to use a thoughtful, scientifically rigorous prioritization process. Period. And as long as people are considering climate change in their decision-making, I think that is a huge step forward. Could it be done better? Yes. Are there other ways of slicing and dicing it? Yes. Does it mean you throw it out and start from scratch? No. I think any effort that generates financing and support for coral reefs is a good thing."

Our food arrived, plates of vegetable-laden omelets and hash browns. I asked Helen the same kind of question I'd asked Alicia in Sulawesi. "Why did Paul Allen and Vulcan fund this?"

Helen said that Paul Allen, who had the means to travel anywhere in the world, loved diving in the Maldives. But for one planned trip, he was told there was nowhere to dive; all the corals were bleached. After that, Paul Allen sought out coral experts to understand what was happening, including the University of Hawai`i's Ruth Gates, who had been a source of inspiration for so many in the coral world before her death just before the Reef Futures meeting. Knowing about Paul's previous interest in space with SpaceX, Ruth met with Andrew Zolli of Planet, the company that makes shoebox-sized satellites, and they developed and pitched the idea for the high-frequency, high-resolution map of coral to Vulcan. The Coral Atlas was born from those connections.

Lizzie said, "Ruth was an eternal optimist. She truly made us believe we could succeed." Both Lizzie and Helen had worked with her in Hawai`i and were greatly affected when she died.

Helen looked down at her plate. "Ruth passed away just two weeks after Paul Allen died of non-Hodgkin's lymphoma." She crumpled up her napkin. "October 2018 was really crappy."

Amid the clamor of the restaurant, a quiet reverence fell into our conversation.

Helen continued. "The analogy [for protecting coral reefs] is sometimes made to rearranging deck chairs on the *Titanic*. But working on the Coral Atlas, my thinking has shifted. Rearranging deck chairs was not helpful. But Unsinkable Molly Brown, she was actually doing things that helped. She saved lives by her actions."

Helen was referring to the story of Margaret "Molly" Brown, a Denver socialite who spent a few months visiting her daughter and traveling in Europe in 1912. She booked passage home to New York on the RMS *Titanic*. After the ship struck the iceberg, Molly helped hundreds of passengers climb into lifeboats before finally being ordered to climb in one herself. Even inside the lifeboat, she urged the officers to search for people swimming in the seas and pull them out. When the officers refused to go near the sinking ship because they feared the lifeboat would be sucked down as it sank, Molly took an oar herself so she could save more people.

"We have to ask, what are the actions that we can take that will make this calamity less bad?" Helen said. "How do we get reefs in 2050 to be in better shape than they would be otherwise?"

Lizzie added, "We know that there will be pockets that survive. We are trying to ensure that in those pockets where they survive, we have the conditions for them to thrive. And then once we reduce emissions, the pockets will provide sources of larvae that can help reseed the reefs. That's the pivot that's happened. How do we put our effort into . . ."

". . . the lifeboats." Helen finished the sentence for her.

THE WEEK BEFORE I VISITED WASHINGTON, D.C., was one of the most difficult of my life. A neurologist we had taken Isy to see had prescribed a more aggressive treatment—an infusion of immunoglobulins, sticky proteins that act as antibody mops. We hoped the process would flood Isy's immune system and wash away the excess antibodies that we thought were attacking her brain. An IV was inserted into Isy's arm as she pressed her face into my rib cage. For four days I sat beside her as the IV pump clicked behind us, dripping into her veins what we hoped would be relief from her mental illness.

The closest experience I'd had to those four days was being on scientific research cruises: the feeling of working under stress, of time being set by the machines pumping fluids, of a focus sharpened, of the precision of measurements, of the unyielding boredom of repetition, of fearing a catastrophic mistake, of not having space for normality, of things like eating and sleeping falling away to the demands of completing tasks. I never kept a journal on a cruise, and I couldn't write during Isy's infusions, either.

Instead of a chief scientist, our expedition into the seas of infusions was overseen by the kindest of nurses, who cradled us in competence and support. At each day's end, she handed Isy a package tied with a colorful ribbon. Inside was a Heath bar, Isy's favorite, with a pink sticky note on it. The fourth and last one said: *So proud of you this week! Tomorrow is a new day! How can you make a difference in someone else's life?* Sometimes, we need to be the ones helping others into the lifeboats. But sometimes, we need help climbing in ourselves.

34

Reef Investments

While philanthropic donations like those of Bloomberg and Paul Allen are significant, they have limitations. Recall that only 0.56 percent of all philanthropy is awarded to the oceans. So, besides meeting Helen to learn about the Allen Coral Atlas, I'd wanted to go to Washington to visit The Nature Conservancy and try to understand their work developing three new financial tools that benefit coral: debt-for-reef swaps, blue bonds, and insuring coral reefs. To my ear, the words *debt*, *bonds*, and *insurance* hold all the allure of saltine crackers. But after talking with people involved in these projects, I felt like I'd substituted rosemary sea-salt flatbread for what I'd assumed were bland bits of the economic system. Whether or not these financial innovations can scale up fast enough to matter remains to be seen.

"One project that I've really come to appreciate," Lizzie had said the first time we met at her house in Austin, "is the debt swap." I nodded my head, acting like I had any idea what that meant, and then asked if she would be able to put me in touch with some of the people involved. She connected me with Rob Weary, who spent two decades brokering those kinds of deals as TNC's deputy managing director of blue bonds.

As background, Rob explained when I spoke with him, in 1998 the United States passed the Tropical Forest Conservation Act. This legislation allows the United States to work with developing countries to exchange their debt for conserving and managing forest habitat. Here's how it works: Suppose a developing country has $400,000 in debt to the U.S. government. The United States agrees to discount that debt and allows a nongovernmental organization like The Nature Conservancy to buy that debt for a fraction of the capital (say $40,000 in this case), and the rest comes from the act itself, basically through appropriations from Congress. The debt is then forgiven, but the developing country provides the full amount of $400,000 in the local currency to local conservation groups to protect and manage forest habitat.

It's a win-win-win: The Nature Conservancy receives $400,000 in forest protection for a $40,000 investment. That's a 10:1 return. The developing country spends the $400,000 for a much-needed investment on its own forests, and it gets to do it in its own currency rather than having to exchange funds into U.S. dollars. And the United States receives some fraction of the money it is owed, but more important, generates $400,000 of conservation activity on the ground. Debt swaps through the Tropical Forest Conservation Act proved successful. More than sixty-seven million acres of tropical forest were saved in fourteen countries. The swaps themselves generated $339 million to protect and manage those forests and leveraged an additional $105 million in financing.

Around 2012, Rob gave a talk on debt swaps, and Ronald Jumeau, the ambassador to the UN for the country of Seychelles, was in the audience. If Africa looked east into the Indian Ocean and Kenya were its eye, it would gaze upon the 115 islands that make up the nation of Seychelles. As an archipelago, it has a coastal economic zone three times as big as Germany. More than two thirds of the country's economy depends on the ocean, for both tourism and fishing. Mr. Jumeau asked Rob if it would be possible to structure debt swaps in the coastal area surrounding Seychelles. Although no one had done any of these sorts of "debt-for-nature" deals in marine systems, there was no theoretical reason it couldn't work.

A few months later, at the United Nations Conference on Sustainable Development in Rio, Rob heard Danny Faure, then vice president (now president) of Seychelles, say publicly that if a debt swap could be arranged,

Seychelles would commit to protecting 30 percent of its coastal waters. What's more, half of that area would be designated at the highest level: no-take marine preserves, where no removal of any marine life would be allowed. "That was a great commitment," Rob said.

A few years earlier, Seychelles had been caught up in the global financial collapse connected to the real estate bubble. "Their bank was Lehman Brothers," Rob said. "And when Lehman Brothers went belly-up, Seychelles did, too." The country was forced into bankruptcy as a result. By the time Rob got involved, Seychelles had restructured much of its debt and was on the way back to financial health. But a $22 million piece of debt was still held by several European countries. A deal for that debt would have to be negotiated through an entity called the Paris Club, which includes officials from all European countries. It took three years of negotiation, but eventually the Paris Club reduced the debt on the $22 million by $1.4 million in exchange for commitments by Seychelles to create the marine protected area. Using a combination of philanthropic donations and their own investment arm, The Nature Conservancy bought the debt and put it into a trust. Seychelles will repay the debt to the trust over twenty years, in its own currency. The trust will disburse funds to support the marine protected area, including an endowment for future programming. Rob said, "It was the first time that a debt swap like this for conservation had ever been negotiated in the Paris Club; first one ever done with impact capital; first one ever done for marine conservation; first one ever done that had the policy commitments of expanding protected areas. So there were a lot of firsts."

In 2018, Seychelles set yet another first. It worked with the World Bank to structure the first bond for marine conservation, something that's come to be known as a "blue bond." Like the debt swaps, these financial tools are modeled on land-based versions called "green bonds." Green bonds are usually issued by governments that need to raise money for an environmental project like a hydropower dam or a water treatment plant. The money investors receive as interest when the bond is repaid is often tax-free, to encourage investment. Green bonds blossomed after the 2016 Paris agreement, reaching $258 billion in 2019, a 51 percent increase over 2018.

Seychelles' bond was issued for $15 million and purchased by three large investors. The funds will be used to manage the country's new marine protected area, as well as to restore coral reefs that have suffered from blast fishing and bleaching. While the bond is the first of its kind, The Nature Conservancy hopes it will be far from the last. It is aiming to use Seychelles' model, combining debt swaps with blue bonds to unlock investment on par with the billions generated by green bonds. Rob has identified four million square kilometers of ocean, including 13 percent of the world's coral reefs, that could be supported this way.

I asked, "So someone like me, as a consumer, would I be able to invest in blue bonds to fund my kids' college tuition?"

"I think that they would eventually get to the point where they would be available," Rob said. "You'd be able to probably buy them in a mutual fund, for example. But we're not quite there yet. But I think they could be there within five years."

On March 26, 2020, Seychelles president Danny Faure made good on his pledge to protect 30 percent of its coastal waters. He signed a decree establishing thirteen marine protected areas, a total area of 154,000 square miles, about the size of California. Half of those waters are entirely no-take. Included in the preserve is the Aldabra Atoll, the world's second-largest atoll, home to the planet's largest population of tortoises and the coco-de-mer palm, which produces the largest seed of any plant. Among Aldabra's coral reefs swim more than a thousand species of fish, humpback whales, and the Indian Ocean's only population of dugong.

WITH TNC'S OFFICES PARTIALLY CLOSED IN FEBRUARY 2020, I sat at an outdoor table at a Starbucks with Mark Way, TNC's director of global coastal risk and resilience, to talk about insurance. Mark had begun his career working at Swiss Re, one of the world's largest providers of insurance and reinsurance. When it launched an initiative to study social, environmental, and ethical risks and opportunities, he became fascinated with the potential for insurance as a tool that could help manage the financial impacts of a more severe climate. Coastal wetlands, for example, save the insurance industry $52 billion through reduced storm flooding, while at the same time both creating income for fishermen by acting as a nursery for young fish and sequestering carbon in living plants.

About twelve years ago, Mark said he was chatting with a colleague, Kathy McLeod (no relation to Lizzie Mcleod), at an industry meeting, bemoaning the lack of innovation from some corporations when it came to sustainability. They wondered whether it might be possible to directly insure nature, for example a coral reef. Some years later Kathy joined The Nature Conservancy and was working with a team in Mexico off the coast of Puerto Morelos near Cancún. The region is highly vulnerable to hurricanes, which could damage the coral. The Nature Conservancy used models to compare the value of having the reef in place to what that reef would be worth if its height were decreased by 1 meter all along its horizontal extent. A healthy coral reef could absorb 97 percent of wave energy as it barrels toward land.

"So then you basically see the delta between the damage you would incur if you lost some of that reef and the damage that you would avoid by keeping it healthy and in place," said Mark. By putting a dollar value on the reef, discussions with local stakeholders become easier. "It is normal to use insurance to protect valuable assets, so why not critically valuable natural assets like reefs too?" he said.

The government set up a trust fund to maintain the reef and beach with the support of TNC. Swiss Re was also a key stakeholder. In June 2019 the insurance policy was issued, the first ever to protect a marine ecosystem. It covered 160 kilometers of reef and beach and would pay out to repair the reef if windspeeds over 100 knots were recorded in a certain area. This avoided the need to assess damage and, critically for the reef, provided a very quick payout—taking just a matter of days. Mark said, "What the insurance is designed to do is to get the money to the policyholder very fast. Because the sooner you can get onto the reef and start the recovery process by, for example, clearing it of debris, it increases the chances of a better recovery. You can imagine that if you've got debris on the reef, and it's rolling around, it's just going to cause more damage." A critical part of the payout was that it would fund the work of "reef first responders," people trained to go onto the reef after a major event, assess damage, remove debris, and make repairs. These teams became known as the reef brigades.

Like debt swaps and blue bonds, this reef insurance policy was hopefully just the start. Mark's program was already working with the Mesoamerican reef, which spans Guatemala, Honduras, Belize, and another part of Mexico. The goal was to be able to deploy reef brigades over four hundred miles of coastline in the wake of storms. And while still in its early stages, there were possibilities that the insurance model would spread to Indonesia, the Philippines, Fiji, and Micronesia, as well as Florida and Hawai`i. But each place has its own challenges. For example, Florida has a risk from hurricanes, but are its reefs already so degraded that using funds to restore reefs is more important than insuring what little still exists? In Hawai`i, the reefs are in better shape, but the major risks are from bleaching and sedimentation, not hurricanes.

The 2020 hurricane season was the most active on record. After naming storms with all the standard letters of the Roman alphabet, the weather service worked through nine letters of the Greek alphabet to name the rest of the thirty storms. When Hurricane Delta hit Quintana Roo on October 7, it brought windspeeds of 110 miles per hour and triggered the payout in the insurance policy. As soon as the weather calmed, the reef brigades deployed, spending hours underwater righting coral colonies and cementing them in place. There were setbacks. The COVID-19 pandemic precluded brigade members from diving every day; the payout didn't reach Quintana Roo as quickly as intended; a second, but weaker, Hurricane Zeta pounded the reef eleven days later, interrupting repair efforts. Nonetheless, by the end of December the brigade had righted and stabilized two thousand large colonies and secured 12,500 fragments of coral to the reef.

WHEN NOBEL PRIZE—WINNING ECONOMIST LIN OSTROM looked around the world for examples of management plans that defied Garrett Hardin's notion of the Tragedy of the Commons, she discovered that they all shared certain traits. Those that succeeded were not entirely private and also not entirely controlled by a central government. A mixture that varied in time and scope depending on circumstances, which Ostrom called polycentric governance, led to outcomes in which common resources were sustained rather than being depleted. Today, a term used for such systems is "public-private partnership." The debt swaps, blue bonds, and reef insurance that I heard about in Washington, D.C., all fell into this category. They involved cooperation by government, nonprofit, and for-profit institutions, as well as the people who live and work near the resource itself. And while these partnerships are still rather novel, they hold the promise of providing much-needed funding to one of the sea's most critical ecosystems at its most critical moment. Whether these funding ideas are "crazy" or scalable remains to be seen.

Later, thinking back on my conversations with Rob and Mark, two comments stood out. Describing what it took to get the Paris Club to come to the table with Seychelles, Rob had said, "It's kind of like, 'Let's get everyone in the room and talk about how to do this with kinder terms.' The agreements had to suit everyone and had to benefit the environment." And Mark had spoken about the challenge of addressing societal goals and business goals, saying, "If you can make them interlink, you're going to solve problems a lot quicker." This kind of work of building connection is the ultimate flaw Lin Ostrom discovered in Garrett Hardin's Tragedy of the Commons: By speaking to one another and working together, we can find ways to serve all our goals.

Just as the coral and the algae have done for so long.

PART VIII

Australia,
from Afar

35

Cancellation

The first time I heard the term "cloud brightening," I was sitting in a huge white tent on a steamy day at the Texas Book Festival. I was part of a panel with Peter Wadhams, a climate scientist who spent decades studying the icy polar realms and had written a book called *A Farewell to Ice*. Peter was an emeritus professor, gray-haired and gray-bearded, with an English accent and a gentle, halting way of speaking that lent a gravitas to his words. He talked about the potential catastrophic runaway impacts of warming at the poles from climate change. What he was really worried about, he said, was the melting of huge reservoirs of frozen methane (charmingly pronounced "meethane") that are currently sequestered below the ice. Because methane has the capacity to hold ten times the heat of carbon dioxide, once the methane starts leaching into the atmosphere, climate change will accelerate so fast that it will be irreversible.

I remember wishing Peter hadn't brought up methane. Methane is like the itch that always sat in the back of the climate science part of my mind, demanding attention. But I usually ignore it because of the rash of carbon dioxide everywhere else. And because thinking about runaway methane is truly frightening.

Peter said that despite the looming reality of climate change, it wasn't all hopeless. Technological solutions existed. These ideas were known as geoengineering.

I winced again, internally, at the mention of geoengineering, which I had always thought of as a last resort, something to try when all else failed, and I was glad when one of the follow-up questions from the audience was about "unintended consequences," pointing out that when we had tried to fix our environmental mistakes in the past, we'd often made things worse.

Peter said he'd studied many different geoengineering ideas. These included shooting sulfates into the stratosphere to reflect the sun's rays, or adding fertilizers to the ocean to stimulate photosynthesis, which uses up carbon dioxide. After much research, he'd become convinced that one idea was worth a try: marine cloud brightening. And he'd come to that conclusion precisely because it was a form of geoengineering for which unintended consequences, should there be any, could be limited.

Just what is marine cloud brightening? Peter explained that it is simply spraying seawater into the air through nozzles that turn the seawater into very, very small droplets, so small they float in the air. Those droplets drift "into the bottoms of gray clouds, like you get in Britain, and it makes them white, and it also reflects more radiation." Seawater is harmless to the atmosphere—crashing waves naturally loft sea spray skyward all the time. But most important, "as soon as you stop pumping through these nozzles, the effect stops. So you can minimize the unintended consequences. And that, I think, is the least harmful form of geoengineering."

Though I had become more intrigued, I still wasn't convinced. As Peter and I sat at the book-signing table after our panel, I asked for more detail. He flipped to the center of his book, where glossy pages held graphs and photos of a warming planet. He pointed to a drawing of a kind of pontoon boat with specialized towers that looked like chimneys, but instead of smoke, a fine mist of sea spray billowed out. Peter said that in the clouds, water droplets gathered around the sea salt in a process called nucleation. Marine clouds naturally have fewer nucleators than land-based clouds, so they are a bit less bright. Adding more sea salt makes them brighter. You can see this same effect by comparing a glass of skim milk to whole milk— the skim milk looks a little bit gray. It's not because of any gray coloring, but because the skim milk has less fat than whole milk. In a water-based liquid like milk, fat forms tiny globules. Those minuscule spheres scatter light. Whole milk has more fat globules than skim milk, so it scatters more light and looks brighter. It only takes a bit more sea salt to get a meaningful effect, Peter said. Add about 1 percent more nucleators to marine clouds, and you could counteract the temperature increase due to carbon dioxide in the atmosphere. "You'd need a flotilla of about six hundred of these vehicles," he said, pointing to the rendering in his book.

Six hundred seemed a paltry number to cool the entire planet. "And how much would that cost?" I blurted out.

"Just around $1.5 billion."

"You're kidding! We could control the planet's temperature for $1.5 billion?"

Peter nodded. But it was a sad nod, a nod of someone who knew that the kinds of serious conversations we should be having about climate change were not yet happening.

ONCE I'D HEARD ABOUT MARINE CLOUD BRIGHTENING, I couldn't stop thinking about it. Was it really a viable solution? Could we—and should we—geoengineer our way out of climate change? I found a long discussion among economists and climate scientists published by *New York Times* climate columnist Andrew Revkin in the online magazine *Medium.* Much of the discussion was theoretical, lots of "what if" and "we can't be confident" and "we must avoid surprises." The problem was that no one had done any real-life experiments, so all the speculation was based on computer models. Ultimately, the discussion focused on human behavior. Several worried that a technological "out" would mean we'd stop dealing with the underlying problem of carbon emissions. It would be like taking cholesterol medicine for vascular disease, but not doing the work of shifting to a lower-fat diet. On the other hand, using another medical analogy, NYU's Gernot Wagner suggested that geoengineering might prompt change: "Seeing what it's like to go through chemotherapy first-hand may well motivate you to stop smoking." None of the arguments were specifically focused on the ocean. (And all omitted the objection that the side effect of ocean acidification from the accumulation of carbon dioxide in the atmosphere would continue.) Amid these admittedly relevant points, I couldn't shake the fact that all of the people in this conversation were white and male and born in the twentieth century.

I asked my son, Ben, who was then sixteen, "Would you be okay with a solution like cloud brightening that slowed the heating of the planet but allowed the fossil fuel industry to continue to pollute, and that might shift people's attention away from the real problems with carbon dioxide?"

Ben gave me the look of a kid asked a question by an adult that wasn't really a question because it had only one answer. "If it's going to stop climate change, why wouldn't we try it?"

Whenever I had a chance over the next couple of days, I posed the same question to Ben's friends. They all responded in the same way. These teenagers recognized that the future they are inheriting is already changed, and not for the better. They don't have patience, or time, for "being confident," placing blame, or assigning responsibility, at least not now. They care about solutions that might ease the future struggles we are handing them.

Ken Caldiera, one of the climate scientists in the conversation Andrew Revkin posted, included a clip from an old email he'd written:

> Economists estimate that it might cost something like 2% of our GDP to convert our energy system into one that does not use the atmosphere as a waste dump. When we burn fossil fuels and release the CO_2 into the atmosphere, we are saying "I am willing to impose tremendous climate risk on future generations living throughout the world, so that I personally can be 2% richer today."

As Greta Thunberg and tens of thousands showed us in the summer of 2019, teenagers have already heard that message loud and clear. And the ones I spoke to believed they deserved to know what the options were, even if these had intimidating names like geoengineering.

AFTER THE REEF FUTURES MEETING IN FLORIDA, I followed up with Dan Harrison about the Australian cloud-brightening program. Dan wrote that he'd received funding to build a one-tenth scale model cloud brightener and attached a rendering of the instrument to the email. It was not the smokestack-looking structure that Peter Wadhams had shown me. Dan had done the energy calculations, and that system was incapable of producing enough power. Instead, the model Dan proposed looked like a wide, short cannon, or like an artificial-snow blower. The end of the cannon was fitted with an array of ten struts, like the spokes on a wheel. Each strut held ten nozzles, making one hundred in all. High-pressure air and seawater were forced out of the nozzles, forming a spray. The key was to produce droplets small enough to float to the clouds about twelve hundred feet high. The ideal size for such buoyancy was around a micrometer, or the size of a typical bacterium. A company in Italy had been contracted to build the cannon part, and a company in China was contracted to build the air compressor. Dan and his team would assemble the cloud brightener in Australia and test it at a research site on land throughout the fall. The first demonstration on the Great Barrier Reef would be at the end of March 2020. I asked Dan if I could be there to see it, and he generously invited me along.

IN EARLY MARCH, I sat next to Isy as an IV snaked into her arm once again, during a second round of infusions. We were watching the seventh and last season of *Gilmore Girls*, a show that had distracted and entertained both of us through some of the worst periods of her OCD. I'd grown fond of the main character, Lorelai, and her daughter, Rory, but I could tell that the rest of the episode wouldn't result in any new revelations. As we watched, my thoughts drifted to the next month's plans, making lists in my mind. Tomorrow, this infusion would end. In two weeks, it would be spring break. We would fly to Chicago and drive east to Indiana so Ben could visit the college where he'd been admitted. Then we'd all pile into the car and drive two hours in the other direction, to Wisconsin, where we would drop Isy off at the residential OCD program where she'd been admitted.

I knew Isy had been tolerating this infusion treatment—days of sitting with the clicking of a pump behind her and needles poking into her veins—but didn't really believe it would work. The first one hadn't. I still held out hope that the immunoglobulins coursing through her body would wash away the anxiety and rules that imprisoned her. Or perhaps just loosen things up so that in Wisconsin she'd be able to do the work of battling her OCD with a bit more ease. I wasn't ready to picture what it might feel like to say goodbye once we got there.

Afterward, I'd have a day to repack, put together my scuba and report-ing gear, and head to Australia. Just the day before, Dan Harrison had emailed me the sampling and analysis plan for the fieldwork. I would meet him in Townsville, which is where most of the academic and re-search institutions that focused on the reef were located. We'd spend six days on a government-operated research vessel called the RV *Jubilee*, sleeping in one set of modified shipping containers bolted to the deck and showering in another. He advised bringing my own sleeping bag. The food would be cold sandwiches and noodles warmed in a microwave. This would not be a cushy trip, but being invited felt like winning the lottery to me.

The aim of the fieldwork was to measure the plume of sea spray pro-duced by the demonstration-scale cloud brightener. To do that, the RV *Jubilee* would direct itself into the wind and cut its motors. The cloud-brightening cannon pointed off its stern would pump water from a few meters below the surface through its nozzles, creating a plume of sea spray. A second, smaller ship and a drone, both equipped with instru-ments for measuring aerosols, would crisscross the path of the plume to characterize its expanse and the array of sizes of the droplets produced. A lot would have to go right for the team to capture the data they wanted: All the vessels would need to be coordinated, data-recording instruments couldn't fail, filters couldn't jam, and fans couldn't overheat. Calm winds and low waves would be critical. To add a bit more to the list of things that could go wrong, this was monsoon season and cyclones could form at any time.

When I saw the map of the locations for the experiments, I couldn't believe that I'd been there before. We'd be adjacent to Davies Reef. Of the fourteen-hundred-mile expanse of the Great Barrier Reef and the three thousand coral reefs there, I was headed back to the very same reef where I had worked in 1992 as a graduate student. Back then, I had helped collect the coral-eating Crown-of-thorns starfish that were at the time the reef's greatest threat (and remain problematic). About the size of a dinner plate, with eight to twenty arms splayed around the central disk like a sunflower, these thorny echinoderms can destroy living coral like a plague of locusts set upon crops. Half the reef bears their scars. Now, I would return to learn about a response to the greatest threat to today's reef: climate change. I would be present for one of the world's first field demonstrations of geoengineering, one that had the potential to actually work on a regional scale.

On the couch next to me, Isy squirmed. She needed a break. We paused *Gilmore Girls*, and the nurse disconnected Isy's IV. I checked my email on my phone. My daily Google roundup of "coral reef" stories had arrived. It was summer in the Southern Hemisphere, and water temperatures were rising in the Coral Sea—not just rising but reaching levels warmer than had been seen since 1990, more than 1.5°C above average. NOAA's Coral Reef Watch program had issued maps where the reds of high bleaching probability were creeping toward the reef, like a storm system gathering to the east. "Alarm bells are ringing," Lyle Vail, the director of the Australian Museum's Lizard Island Research Station, told *The Guardian*. The story reminded me that I hadn't emailed Dan Harrison my flight information, though he'd requested it earlier. I quickly typed a note, "I arrive in Townsville on Qantas 1792 at 1600 on March 21," and then I thought about it and added, "coronavirus willing." The virus, which was just hitting the news when I went to Washington, D.C., was now spreading, particularly in South Korea and Italy. And while there were reports it was more contagious and deadly than the flu, much remained unclear.

The next day, a new message from Dan arrived. The Australian government was taking COVID-19 seriously. Because of the early outbreak in Italy, the Italian colleague who had helped build the cloud-brightening instrumentation would not be allowed aboard the ship unless he quarantined for fourteen days upon arrival. It was possible such restrictions might apply to me as well. I counted the days. If I switched my flights and left soon, I could quarantine for fourteen days before the ship set sail. But even as I looked at my calendar, I knew I would never do it. I wouldn't leave before taking Isy to Wisconsin.

Five days later, Dan emailed again: "Bad news Juli."

Restrictions had been placed on all government facilities, including vessels, for all international travelers. I would be barred from any of the cloud-brightening fieldwork.

As Isy's admission date grew closer, I stubbornly held on to my plans to go to Australia. A travel magazine had offered me a commission for an article about tourism on the reef, and they still wanted the piece. Airplane surveys of the warmest northern region posted to Twitter showed a reef splattered with the white of coral beginning to bleach, an unfolding story I could still observe. I imagined I might even hire a private boat out of Townsville and watch the cloud-brightening plume from afar.

The week before the start of spring break, the news filled with COVID-19 tragedies in New York and then Boston. Painful footage of overwhelmed, undersupplied hospital workers and first responders filled my screens. We canceled Ben's college visit, which wasn't essential, but mixed messages from government officials made decision-making about my trip to Australia confusing. I held on to my ticket, even as I knew I shouldn't.

Four days before Isy's admission date, Keith and I pulled up a map of the United States and plotted the drive from Austin to Wisconsin. It was eighteen hours each way, but by that point we weren't willing to risk exposure to the virus in an airport. Our only goal was to get Isy safely to the program where she could get help. We packed our bags and loaded them in our car. But still I waited hours before I could bring myself to log on to the airline's website to cancel my trip to Australia. It was almost nothing compared with all the loss that was to come. But at the time, it felt like everything.

36

Collapse

Face masks were in short supply everywhere, and the CDC hadn't yet made clear that they would be effective. But I was convinced that they were one of the few steps we could take to protect ourselves from the virus, which seemed to be airborne. As we drove north out of Texas, I hand-sewed face masks using a pattern I'd downloaded from the internet and old material I'd excavated from the back of the linen closet. The fabrics were leftovers from our family's life: blue and white flowered pillows I made for a porch chair; the train tracks of Ben's little-boy curtains; a turkey pattern from a skirt the kids made for my mother, who gamely wore it one Thanksgiving; the watery jellyfish print that I'd used to make bookmarks to give away at book signings.

We stopped near Dallas and again in Joplin. Whenever Isy got out of the car, she wore the jellyfish-fabric mask. She touched nothing except the toilet handle, and then she washed her hands in the thorough way OCD forced her to. I still had no idea how soon we would all become masked. I still had no idea how we would all come to fear the invisible in the world around us. Looking at her, I shrank from the transformation of daughter to patient.

By the middle of Illinois, it was gray, dreary, spitting rain. The cornfields were cut short. The trees were leafless and skeletal, as if the world felt as bereft as I did. On the radio, I listened as the government began to take the virus seriously. Groups of ten or more people were no longer permitted to gather. The Bay Area instituted the first shelter-in-place order. From our hotel in West Milwaukee, we ordered takeout, and I drove through narrow streets past old buildings with ornate edging to pick it up. Inside the restaurant, paper leprechauns were pinned to the walls and strings of green beads hung from the chandeliers in anticipation of a celebration that wouldn't happen. I had totally forgotten that tomorrow would be St. Patrick's Day. The country was starting its giant time-out.

After dinner, we watched the last episode of *Gilmore Girls*. My tears were close to the surface, and I cried when Lorelai called her mother from the top of a mountain. I cried when Rory told her first boyfriend that he had meant safety and kindness to her. I cried when Lorelai gave Rory her blessing to write the story of their lives.

We got ready for bed, and Isy asked me to tuck her in like when she was young. She hugged me tight—so tight it felt like the way she had clung to me as a toddler. It was hard to fall asleep, hard to imagine that tomorrow my child would be somewhere else, and I wouldn't know when I'd see her again. The treatment facility had said that because of the virus, it was unclear whether we would be allowed to visit during what would be a months-long stay. I looked across the gap in between our beds, and her big green eyes looked back at me, stark against the white sheets. She couldn't fall asleep, either.

WHEN WE ARRIVED AT THE FACILITY, Isy's temperature was taken. It was the first time we were asked the battery of questions that would soon become so familiar: Had we had a cough in the last forty-eight hours? Did we have a fever? Had we been around anyone exposed to the coronavirus? Had we been knowingly exposed to coronavirus? No. No. No. No. We met various staff in a large room, couches situated well over six feet apart from each other. We retold psychological and medical histories. We signed paperwork.

Finally, the nurse said, "It's the moment that you didn't want to come." She moved toward the door. "I'll give you a few minutes."

Isy clung to Keith in the hug she'd given me the night before. She turned to me and started crying into my shoulder. I held her and stroked her back. She said, "I didn't expect to cry." When the nurse returned, Isy pulled away and leaned down to grab her suitcase and water bottle. The bottle was covered in stickers of places she'd visited, evidence of the freedom she was relinquishing.

The nurse said, "I'm sorry, but you can't bring that. We can't guarantee it'll get washed, and we need to keep you safe."

Isy handed the water bottle to me and walked through the door.

Clutching the bottle outside in the bright, cold sun of the afternoon, I felt something crush deep in my sternum. It hollowed out my bronchia and emerged as a high yelp, a primal gasp, a sound like no other I have ever made or heard. I inhaled roughly and exhaled tearless sobs. My cries were dry ones, heaves with nothing to lubricate the pain. Keith pulled me toward him, supporting me. When we reached our car, its trunk was wide open, all our belongings exposed. We'd left it like that all day, our vulnerability on display for all to see.

DURING THE EIGHTEEN-HOUR DRIVE HOME, Keith and I were quiet and gentle with each other. I had few words, and he didn't ask for any. When it was my turn in the passenger seat, I checked my email. My Google alerts told me that in the South Pacific a cyclone had formed and was tracking toward the Great Barrier Reef, potentially mixing deep water and cooling the corals, but that was all I could find. The news was overwhelmed with the rapidly changing response to the coronavirus. There were no more reports on the mass bleaching, the other epidemic beneath the seas.

Another thing I didn't know at the time—I wasn't the only one on a long, cross-country trip, trying to get one more critical thing done before COVID-19 restrictions went into effect. Dan Harrison and his team had packed up their vehicles, with all their scientific gear and the cloud brightener itself. They were making the seventeen-hundred-mile drive from Southern Cross University to Townsville. Dan and his team camped alone and cooked their own food along the way to avoid contact with others. They were trying to get out to sea before everything shut down.

AUSTIN WAS AN ENTIRELY DIFFERENT WORLD from the one we had left four days earlier. Everyone was on lockdown, and I was probably in shock. I couldn't watch the news or any media. Everything that was supposed to be entertaining seemed vapid. Keith scrolled continuously through the stock market on his phone. All along the right side, I saw red numbers rushing by. It looked to me like an accounting of jobs that were going to be lost, the lives that were going to be lost, the struggle the world was headed into. I was linked to email chains from groups of friends and writers. They wrote about what food to eat, how to handle daycare for kids, what to read. At first it was comforting, but as the days rolled by, the collective panic tended toward greater isolation. Caregivers, take-out food, and mail were all suspect. Social distancing turned into social dysfunction. Alarms were going off everywhere.

Alarm is a good thing. We need it to survive. But what happens when alarm goes haywire? You get obsession and you get compulsion. You get OCD. You get frozen. You get admitted to a residential treatment facility. You get corals launching an immune response against their own solar power systems. You get bone-white reefs. You get starvation. You get mass mortality. You get collapse.

IN EARLY APRIL, the scientists in Australia completed their assessment of the damage to the Great Barrier Reef from the record-high February temperatures. Terry Hughes, who surveyed the extent of the bleaching from a small airplane, posted videos: splotched white haloes of bleaching surrounded reef after reef. He wrote in one poignant tweet, "It's been a shitty, exhausting day on the #GreatBarrierReef. I feel like an art lover wandering through the Louvre . . . as it burns to the ground." The 2020 Great Barrier Reef bleaching was the most expansive in history, encompassing a larger area than either 2016, which scarred the northern reef, or 2017, which hit the central section. This event included the southern region, which had yet to experience a bleaching. The worst part was that these three major mass bleaching events had all taken place in the span of just five years. Before that, the most recent mass bleaching was 2002. And before that in 1998. And before that, the Great Barrier Reef had really never experienced one.

But true to its name, the Great Barrier Reef is truly great. Its massive scale, fourteen hundred miles end to end, stretches even farther than Italy top to bottom. It could cover the West Coast of the United States and drip off the edges to Vancouver and Tijuana. Even during this most extensive mass bleaching, many parts remained unaffected. The outer reefs seemed to have been cooled enough by the cyclone and were spared the worst of it. Many of the most popular tourist spots were left unscathed. How was that possible? Each polyp is an amalgam of its own genetics, its own microbiome, its own algal symbionts. And all of those variables work together so that each polyp responds to the individual conditions it finds itself in at any moment. Hundreds or thousands of polyps make up a single coral. Thousands or tens of thousands of corals make up any reef. Thousands of reefs make up the entire ecosystem. Considered that way, the Great Barrier Reef is trillions of individual responses. That's a lot of opportunity for survival.

ISY CALLED HOME EVERY NIGHT. She was adjusting. She was learning a kind of therapy known as exposure and response prevention. She had built a hierarchy of her obsessions and compulsions and ranked them easiest to hardest. There were more than eighty of them. Starting at the bottom of the list, her therapist assigned her exposure drills, which meant putting her in situations to trigger her obsessions and then asking her to resist performing compulsions. She recorded how high her anxiety went on a scale of 1 to 7. She timed how long it took for her anxiety to fall by half. She repeated each exposure five times and watched the evidence accrue that she could more easily tolerate the trigger without performing the compulsions. She was starting to believe that all her OCD could throw at her was anxiety, and she could learn to tolerate the anxiety. Once an exposure was capable of raising her anxiety to only a 2 or 3, she crossed it off the list and added a new exposure, a harder one. Isy spent three hours every morning like this. And another hour and a half in the evening. She was acquiring a new language, words like *ritualizing* and *reassuring* and *anticipatory anxiety*. She was still struggling to do schoolwork. The food wasn't so great. She liked her roommate.

IN THE EARLY WEEKS OF SELF-ISOLATION, my Google calendar, which I hadn't bothered to update when I'd canceled my trip to Australia, sporadically fired off confirmation codes for hotels and flights. They were obnoxious reminders of where I might have been, "what if" distractions from the heartbreaking events happening around the world. Yet there was one event I scanned the news for every day. I'd emailed Dan Harrison a few times for updates with no response. Finally, in mid-April, he sent a link to an exclusive story in *The Guardian*, with a video. Drone footage showed the RV *Jubilee* in the distance sailing over a smooth turquoise sea. The shot shifted to the stern of the vessel, and there, bolted in place, was the cloud brightener, pumping out its sparkling mist. They'd made it out!

Close-ups showed the hundred stainless-steel nozzles atomizing seawater into a fine spray. Only 1 percent of the nozzles' work was visible, Dan said. Most of the particles were far too small to see. And that was the whole idea. Those trillions of invisible, micron-sized bits of saltwater were the ones that would float upward, reaching the clouds, enhancing the sparkle that could help beat back the sun's rays. Measurements by the sampling boat and drone detected the aerosols nearly four miles from the cloud brightener—twice as far as Dan had predicted.

For fifty thousand years the Great Barrier Reef has been home to more than seventy groups of Aboriginal and Torres Strait Islanders. They have extensive experience and knowledge of the marine world, and their culture and spirituality rely on the reef itself. These Traditional Owners work with the Great Barrier Reef Marine Park Authority, who oversee and permit all activities on the reef. As I spent time on the Australian websites, I noticed that in any speech or written material, AIMS first acknowledged the Traditional Owners. Usop Drahm, a Traditional Owner from Mandubarra Sea Country north of Townsville, was part of the cloud-brightening expedition. He said, "As an indigenous person, our seas are really important to us, and it's good to see the scientists start to help the sea."

The experiment was only the first step of many that need to be taken before we know if cloud brightening will work, Dan pointed out. And as Misha Matz wrote when I posted the link to Twitter, "so far they just made salt spray." True, but the fieldwork had proven that the salt spray was exactly the kind they'd wanted to make.

The story ended with a statement by Dan: "If this works as well as we'd hope, it might buy us a couple decades. But at the same time, it's absolutely essential that we reduce our emissions." During the COVID-19 outbreak and the subsequent collapse of the world's economies, carbon emissions fell more than at any other time in history. The greatest decrease occurred in April 2020, when emissions dropped almost 17 percent globally compared to 2019. The cost, however, in the heartbreaking loss of lives, jobs, and income was unacceptably high. As we regroup from this crisis, emissions are predicted to resume, and perhaps even bounce higher. Yet the data tell us that we can make changes to our emissions on a global scale. We will never be able to unsee those numbers.

37

Flicker

Two days before driving Isy to Wisconsin, I woke very early in the morning, full of nerves. As had become my habit, I grabbed my phone off my nightstand to doom scroll through stories of the escalating pandemic. But that morning I read a horrifying story about a different kind of illness, one that's afflicted us for much longer, for hundreds of years.

A medical technician named Breonna Taylor had been shot to death by police in her own home in Louisville, Kentucky. Nothing about the story made sense. The three officers who beat down Breonna's door in the middle of the night had a no-knock warrant. Scared for the couple's safety, Breonna's boyfriend reached for a gun he was licensed to own and fired. The officers sprayed the apartment with dozens of shots, eight of which hit Breonna. Breonna Taylor was Black. All of the officers were white. They were at the wrong address. The person they were looking for lived ten miles away. I scanned for more than just the single story about the atrocity. When I finally found one, a detail jumped out. Breonna had been up late playing video games, like I had so many nights, trying to ward off stress. But she didn't have the chance to wake up and reach for her phone on her nightstand like I had. What made even less sense was that, despite its incomprehensibility, the story faded from the news.

Seventy-three days later, George Floyd was murdered in Minnesota when a police officer kneeled on his neck for eight minutes and forty-six seconds. This time, in part because the heinousness was caught on video, the story did not fade from the news. It seemed that because we'd been in the relative quiet of quarantine, we were finally able to make room for this kind of story. Protests broke out around the world in opposition to the systemic racism that results in Black Americans' dying at the hands of police at twice the rate of white Americans. It's the same systemic racism that results in the percentage of incarcerated Black people being almost three times higher than their percentage of the U.S. population. For Hispanic people, it's almost one and a half times higher.

The day George Floyd died, a white woman in New York's Central Park called the police to report that a Black man was harassing her. He was simply out bird-watching. Twitter lit up in outrage. In response, a herpetologist helped create #BlackBirdersWeek. It went viral; thousands of Black people posted pictures of themselves out in nature, binoculars in hand. That led to a string of hashtags: #BlackHikersWeek, #BlackAstroWeek, #BlackBotanistsWeek, #BlackWomenInSTEMWeek, #BlackNeuroWeek, #BlackAndStem, #BlackInTheIvory, and #BlackAFinSTEM with the aim of countering assumptions that Black people weren't out in nature or successful in science. Two of the hashtags, #ShutDownSTEM and #ShutDownAcademia, evolved into a virtual sit-in. Rather than continuing business as usual, academics were asked to spend the day in education, action, and healing around racism. In Misha Matz's coral lab, he encouraged everyone to take part. At the weekly meeting, we would discuss what we'd learned during the shutdown.

The meeting took the form of other Matz lab meetings, with each person presenting an article in turn over Zoom. The first student read a paper that examined the underrepresentation of Black professors in evolutionary science. A 2017 survey from the National Science Foundation reported that African Americans made up only 3 percent of the professional biological workforce; within evolutionary science, that number was only 0.3 percent. The author, a Black member of the American Association for the Advancement of Science, spelled out a history of institutionalized racism as the underlying reason. Another student presented a paper that focused on the unrecognized efforts of Black professors in academia who bear a disproportionate amount of responsibility for the education of Black and Brown students, serving "as role models, mentors, even surrogate parents to minority students, and to meet every institutional need for ethnic representation." A third presented a study called the Diversity-Innovation Paradox in Science. The authors analyzed almost every PhD dissertation—1.2 million of them—in the United States between 1977 and 2016. Not surprisingly, they found that diversity breeds innovation. At a disproportionately higher rate, minorities made novel innovations in their fields. The paradox was that innovations made by minorities were less impactful than innovations made by people in majority populations. As each paper was presented, it became more and more clear how deep the tendrils of systemic inequality reached.

Throughout the conversation, a PhD student named Kristina Black remained nearly silent. When her turn came to present, Kristina said, "I didn't read anything for today's meeting. But that's because I didn't need to. I've lived it." As an indigenous person working in science, she recognized the times she'd been "the diversity hire." She said this knowledge had made her feel both inadequate and unqualified, but there was something more. Kristina told a story about working in a forestry program. Part of her job was to sample DNA from illegally poached animals. One day a year, forestry scientists were allowed into a secure warehouse where the carcasses were stored to take samples. Everyone there, except her, would be white and male. Her superiors coached her on how to fit in, quizzing her on the names of football players and the previous week's scores. She looked at the screen and asked of the rest of us, "What's the point of diversity if I have to learn to be like you?"

Before the protests of the summer of 2020, I wasn't sure that stories of systemic racism had a place in this book. After that summer, I was certain that I couldn't write this book without including them. Not just because this book focuses on the work of academics, which is steeped in and has a very old legacy of racism, but also because coral reefs themselves are critically affected by these issues. Given their tropical location, reefs are located where many people of color traditionally live. Yet because of the long history of European colonization in those places, they are often managed by white people. Although there are exceptions, tourism, predominately by white people who have the means to travel to the tropics for vacation, often compromises conservation efforts. The declining health of the reef disproportionately affects the local people, usually people of color, robbing them of a significant source of protein and the protection from ever stronger storms and rising seas. Lack of infrastructure, sanitation in particular, exacerbates coral and human disease in these regions.

About a month after #ShutDownSTEM, I read an essay by Cinda P. Scott, who is Black and heads the Center for Tropical Island Biodiversity Studies in Panama. Dr. Scott wrote about being placed in remedial reading in her Massachusetts sixth-grade class for no obvious reason except the color of her skin. She wrote that while she swam across the pool during a scientific diving test, the white instructor bellowed at her to swim faster. When she was too slow, he simply failed her without offering any chance to improve or try again. And while her background and experiences are unique, she wrote about the stark outcome of the accumulation of barriers like these: While marine science admittedly makes up only 0.44 percent of all graduate students in science and engineering, Black women in marine science represent a mere 0.003 percent. Dr. Scott wrote:

> Systemic change is needed urgently in all sectors of society where racism pervades. Academia is no exception and the need for change is high. This is not just a university or departmental issue, it is at its core a matter of people treating other people with dignity and respect, and reversing waves of inequality that have been entrenched in our educational system for centuries. If we can embrace the immense diversity found across thousands of species, then surely, we can learn to accept the beautiful diversity that exists within our own.

AS THE VOICES OF PEOPLE OF COLOR rose in a chorus around the world that summer, as I typed it over and over in connection with coral, I had to consider the word *bleach* itself, one also imbued with a sense of color, although, of course, referring to its absence. *Bleach* is not of Latin or Greek origin like so many words in science, but rather Germanic. It comes from *blaec*, which is related to the word *bleak*. As an adjective, *bleak* means bare, exposed, and desolate. Related to a situation, *bleak* means unpromising, unfavorable, and grim. Those words well describe fields of coral stripped clean of their algae, bones dry of their flesh, skeletal remains, and the future of corals in the face of the systemic challenges posed by our future seas: polluted water, destructive fishing, and the ever-approaching thermal limits. And the words *unpromising, unfavorable,* and *grim* are also apt descriptions of all that is true for the hundreds of years of systemic injustices that have become ingrained in our society. They are also words that characterize the future for so many who will contend with the worst that climate change has in store if we don't address it. The people affected will be disproportionately people of color. Coral bleaching is visible evidence of climate change and our dependence on fossil fuels—problems in no way disconnected from the Black Lives Matter movement. Ocean justice is inextricably bound to social justice.

During Isy's stay in residential treatment, the doctors said they weren't sure Isy had PANDAS. And maybe she didn't. But what is a mental disorder if not an autoimmune disease, the body turning against itself? What is racism if not humanity turning against itself? What is climate change if not our behaviors turning against our own well-being? We have so little knowledge of the chemicals that swirl through our brain and make us think the thoughts we think. The surge of oxytocin that warms our heart. The little jolts of dopamine that elevate our thoughts and make them into entire narratives. Or the shots of serotonin that break our thoughts to bits, smashing conclusions like pieces of broken ceramic. What mixture of chemicals fire in my head when I watch my kid fall to pieces? What synapses drive my heart into my chest and tears into my eyes when I read dreadful headlines? What shifts the power away from my frontal lobe to the fight or flight of my basal ganglia when I see someone coughing without a mask? We know there's a love hormone, but we have never identified a hate hormone. Instead, hate must be taught. The strings of neurons must be formed and connected through upbringing, pain, or fear. There's no burst of dopamine in response to hate unless we train our brains to make one. Why did we evolve in such a way that we come equipped with the chemistry for love, but we must rig up the neural lines for hate?

Digging even deeper into the etymology of the word *bleach* through *bleak*, I found another, somewhat obscure meaning. As a noun, rather than an adjective, *bleak* is a small silvery shoaling fish of the minnow family. I imagine that the fish was given the name for its shimmering, bleached-looking appearance. The idea makes me think of the flashes of light that bounce off the scales of small schooling fish. It conjures flickering hope lapping at the shores of a more just and resilient world.

38

Survival Genes

As summer arrived, Texas became a hot spot for COVID-19. The number of cases grew, and I found myself shrinking. Not just from public interactions, which I'd already grown used to avoiding, but from my own life. For more hours than I want to admit, I played a video game on my phone that involved the story of Lily, her deadbeat boyfriend, and her mysterious great-aunt Mary who'd willed her a mansion and a garden. The game was structured so that I could only read the next installment of Lily's story after blasting dandelions to pieces, exterminating garden gnomes, and matching the colors of mushrooms with little square blocks. I tried putting my phone out of reach so I couldn't open the app. And then, when anyone called, it was an excuse to reach for the phone and play. I could focus on the ordered little pieces falling in predictable ways when everything around me felt like it was falling to pieces. It made me think just hard enough to block out the mental noise of the news and the worry about Isy. I could melt into the underlying theme of restoration, and the virtual thrill of ripping out weeds and designing flower beds with just a tap of the fingers. But most of all, when I failed, I always got another chance. Even if I used up all my lives, all I had to do was wait until tomorrow to try again.

Writing can be done in isolation, and I was grateful that my work spared me the exposures to the coronavirus so many others couldn't avoid. But like the heat-shriveled leaves on the bushes outside, my words became desiccated. Instead of writing, I read. I read the stories of heartbreak, of families decimated, of ICU nurses and doctors overwhelmed, of more death and loss than I had ever known. I read how the virus was unpredictable—that it attacked the lungs in some, the heart in others, while still others suffered from kidney damage. Survival depended on blood type, on preexisting conditions, on socioeconomic differences. I read of the disproportionate infection and suffering of people of color and minority groups. I read in the medical science of the vast array of responses to the virus, the different ways of testing for the virus, the efficacies of different treatments. I read until I couldn't read, and then I turned back to Lily and the achievable task of restoring her digital garden.

In between, I thought of the corals of the Great Barrier Reef working to recover from their own pandemic beneath the seas. I spoke a few times to Master Reef Guide Eric Fisher, who lives and works in Cairns, a major access point to the northern reef. Master Reef Guides undergo extensive training through the Great Barrier Reef Marine Park Authority in order to provide tourists with historical and biological context during their visit. Eric had an undergraduate degree in marine biology and was working on his PhD on fish behavior, teasing out where and when different fish gather to eat or to reproduce.

Eric said that while the 2020 bleaching had been extensive, it wasn't complete. "You've got a reef that could be totally bleached and really suffering, with really high mortality, and the next-door-neighbor reef has recovered quite well, or didn't even bleach at all." Patchiness, he said, was what characterized the reef. Each polyp, each colony, each reef responded differently. This is what unifies life on our planet: variety. And the source of that variety is the long string of molecules that make up genes. We are all—from the coral waving its tentacles on the Great Barrier Reef to Dr. Anthony Fauci giving yet one more interview—an assortment of genes, and, through those genes, an array of responses. Even the novel coronavirus itself was nothing more than a string of molecules capable of mutation. That summer, I read about an early variant discovered in India—two beads on the string of molecules were traded for others; these were random mutations. Those first reported changes made the virus even more contagious but, fortunately, not more deadly.

Along with the cloud-brightening cruise and bearing witness to the extensive mass bleaching, one more reason for my trip to Australia was to visit the National Sea Simulator, colloquially SeaSim. Less than a decade old, SeaSim is housed in a half-acre building and has a holding capacity of 3.65 million liters of seawater and thirty miles of pipework. SeaSim's purpose is for the exploration of variability in the genetics of the Great Barrier Reef's life, especially its corals.

For those unable to visit, you can take a virtual tour of SeaSim online. The camera pans across the wide-open space of the main lab, showing rows and rows of aquaria lit with high-powered lamps. Gleaming freezers and massive filtration systems adorn the walls. Clicking on one gray-outlined door leads to a coral nursery full of still more tanks and filters. Other doors lead to experimental rooms, darkened and lit in blue light. Another door opens to a laboratory with chemistry benches extending down either side of the room, and white lab coats hanging on hooks near the door. Exiting the lab, you find yourself on a massive outdoor patio shielded from the sun by blue plexiglass. It too is full of tanks and pipes snaking overhead. All this sophisticated seawater circuitry provides precise control over light, temperature, pH, salinity, sedimentation, and pollution so that the scientists can determine how each affects the health of coral. SeaSim is probably the most advanced aquarium in the world.

Misha had a years-long collaboration with Line Bay, the scientist at the Reef Futures meeting who had shown the giant mind-map of what it would take to breed corals for a future Great Barrier Reef, and put me in touch with her. She had offered to give me a tour of SeaSim, where she oversees the coral genetics, while I was in Australia. Instead, we met by Skype. Due to the time difference, she drank a cup of morning coffee while I contemplated pouring a glass of wine. With her blond bob and stylish glasses, Line spoke with a measured yet warm manner that encapsulated the seriousness of her position and the optimism required to do her work. Laying out the range of research questions under her purview, she said, "Can we identify genetic markers that underpin heat and tolerance? That's one component of it. And the other component of our program is aquaculture, breeding of corals and delivery. Can you promote the spread of the existing variation and in that way facilitate adaptation at a reef level?"

And, she emphasized, those scientific considerations could not be seen in isolation from the cultural value of the reef. As with the cloud-brightening work, Traditional Owners are involved in decisions that surround the work at SeaSim. "Indigenous peoples have a long relationship with various reefs and as part of their culture and belief system," Line said, and they are not comfortable with coral being removed from the reef unnecessarily. "So if we can identify the corals that are at the extremely heat-tolerant end of the curve and multiply those in the lab, that actually overcomes that cultural challenge. There are multiple reasons for wanting those genetic markers."

More than a decade ago, Line and Misha Matz teamed up to investigate some of the most basic, but critical, questions about coral genetics, beginning with: Can corals pass heat tolerance on to their offspring? At the time, there was no SeaSim facility full of replicate tanks with precise controls. Instead, Line said with a smile, "The work Misha and I did were kind of in lunch boxes." Using experiments like the ones Gregor Mendel performed with his tall and short pea plants back in the 1880s, they crossed corals from warmer regions with each other; corals from cooler regions with each other; and corals from warmer regions with those from cooler regions. They exposed the resulting larvae to increasingly warmer water and monitored which ones survived best. Larvae with two warm-water parents were ten times more likely to survive heat than the offspring of cool-water parents. Even having just one parent from warmer waters helped: a mother from warmer waters increased a larva's chances about fivefold; a father about double. (One potential reason for the discrepancy is that the egg is packed with the proteins needed for a larva to survive its initial days, while the sperm isn't, making the mother's contribution larger.) There were two big takeaways: that thermal tolerance is inherited, and that corals have a lot of genetic variability.

The next step was for Line and Misha to try to figure out which genes were responsible for that heat tolerance. Luckily, Line said, they had gotten some help from an unexpected place. "Some collaborators from Columbia University. They are human geneticists, and they came to us and said, 'We love coral reefs. Can we help you?'" She stopped short of telling me the whole story because the paper on that project would be coming out soon in the prestigious journal *Science*. She couldn't talk about it until it was published.

About a month later, I took a break after completing level 384 of Lily's Garden to shoot Misha an email asking if he'd had any word about the paper in *Science* that Line had mentioned. He wrote back saying yes, it had just come out, and agreed to walk me through it via Zoom. He picked up where Line had left off, explaining that the unexpected help came in the form of a geneticist at Columbia University named Zach Fuller, who typically studies diseases in humans. "He just contacted me out of the blue saying, 'We know how to do high-end genomics and GWAS, and we have some extra funds left, and we want to do a coral Genome-Wide Association Study for coral bleaching.'" Misha almost laughed because the offer was so unexpected and generous.

GWAS is one of the most powerful statistics tools in genetics. It has paved the way for identifying many of the genes that we know cause complex human diseases, including the BRCA gene that causes breast cancer and the genes responsible for Crohn's disease, Parkinson's disease, and diabetes. GWAS turns the intuition you might have about how to study genes on its head. Instead of trying to pick which genes help a coral survive better in warmer water and then testing whether your guess is right or not, GWAS considers every gene in the whole genome a potential candidate. Imagine you knew nothing about how beer was made, but you were interested in knowing what makes it more bitter. Comparing many recipes of bitter and not-bitter beer, you'd find similarities in malt and yeast but a lot of variation in hops. You would correctly conclude that a beer's bitterness depends on the amount of hops. Likewise, comparing all the genes from coral that do and don't survive in warm water, GWAS would scan for the biggest differences to find the genes that matter most for surviving warmer water.

Each member of the GWAS team had a specific role. Zach would do the heavy lifting on the sequencing and statistics side up at Columbia. Line, who had collected tissue during the 2017 mass bleaching on the Great Barrier Reef, was responsible for pulling together a lot of samples of coral that survived higher temperatures. Misha's job was building a high-quality reference genome of an average coral. He said, "We needed to have a really nice DNA from the coral: not fragmented, really fresh, really high molecular weight, because we wanted to use all the fancy technologies. It turns out none of the stuff in my freezers was good enough." Luckily, not only was the species they were using, *Acropora millepora*, important to the Great Barrier Reef, it was also well known in the aquarium trade. Crossing the barriers between science and the aquarium industry, Misha reached out to the Austin Reef Club, the same group that had sponsored the coral swap I'd attended in the church gym. One of the members, John Sorkness, was thrilled to donate a few fragments for the study. That humble coral from a tank in Austin, Texas, was the first complete genome of *Acropora millepora* to be published. Decades from now, scientists will still refer to it.

Misha shared his computer screen with me and pointed to a graphic known in genetics as a Manhattan plot because of its similarity to an urban skyline. The horizontal axis was the coral genome laid out chromosome by chromosome. The vertical axis showed the differences in code between bleached and unbleached coral DNA. The higher the dot on the y-axis, the greater the difference between the DNA of the two types of coral. Misha moved the arrow on his computer back and forth across the skyline of the Manhattan plot. "Typically it looks like a bunch of peaks on a smaller landscape," he said. "Here, we really don't have any prominent peaks. I mean these two maybe"—he hovered over two midsize towers—"but even they don't break through. So there are no obvious major genetic determinants for bleaching."

This didn't sound encouraging. "Um, one of the things that Line said was that she was hoping to find genetic markers for bleaching," I said.

Misha shrugged. "There are no specific diagnostic markers. We might be able to fine-tune and sharpen this estimate and get a better idea if we have thousands of corals rather than a couple of hundred, but it's really clear that there are no major determinants, no major signals, nothing like blue-eye gene in humans or lactose intolerance. Almost every gene in the organism has an effect. These effects are tiny. And the whole variation in the population is basically the sum of these super-mega-tiny effects. The whole trait is smeared across the genome in a very thin layer, like butter across bread." In other words, there were no hops in the recipe for corals' tolerance to higher temperatures. Instead, the genes work together in still-mysterious ways to make a coral more—or less—capable of surviving warm waters. Like so much with coral, bleaching was variable, complex, and interconnected. There wasn't just one way a coral could survive; there were untold ways.

Misha scrolled to another figure. "This, I think, is the most interesting result of the paper," he said. The graph showed how different characteristics contributed to coral survival in heated waters. Their genetics were the least important. The most important were the environment where the coral lived and the coral's symbiont type. Remember the idea of solar shuffling in which a coral switches its algal power source for a more heat-tolerant strain? If a coral made a deal with the type of algae called *Durisdinium*, it was more likely to survive, by a lot. *Durisdinium* remains in symbiosis with coral at temperatures a degree or two higher than other strains of algae. But there's a cost: *Durisdinium* shares a smaller fraction of its sugar with the coral.

I asked how these results changed the understanding of the genetics of coral reefs.

"Well, we need to rethink everything," Misha said. "What we can say is that most of the [genetic] variation [for surviving bleaching] is explained by what kind of symbiont you have. So we are back to the nugget-of-hope paper."

He was referring to the last line of the paper, which described the solar shuffle; it suggested that corals could be able to survive in warmer temperatures by shifting solar power sources. *Durisdinium*, even as it forced the coral into conditions of austerity, was a nugget of hope.

I asked Misha where his research was headed next.

"Now we know that the heat tolerance of the coral is probably less important than the whole symbiont interplay," he said. "But the question remains. Is the preference for the symbiont under genetic control? Things become more complicated because there are two players in the game."

If there isn't a strong genetic signal for surviving at higher temperatures, maybe the question isn't about thermal tolerance at all. Maybe it's about having the genes for a merger with the kind of algae that's less likely to bleach at higher temperatures. "And?" I asked.

"Stay with us. The story continues."

AS THE QUARANTINE WORE ON, I thought often of stories and their form. Stories follow waves. They rise and fall, and we are sucked into the energy of their action. When that wave crashes, the story ends and we walk away, satisfied with the ending. While all waves resemble one another in form, like every story, each wave is different. There are none of the same height, breadth, or frequency. There is variation, from the slight ripple to the massive tsunami.

Corals live beneath the waves, witness to the forever motion of the moon's pull and the spinning of the earth. They house myriad stories, which they share through the generations, spun together with the golden rays of the sun and recorded in the layers of stone at their feet. For them, the story's form is not a crescendo falling to completion. It is an endless tale, one that existed before the very notion of story and that will endure well beyond. For coral, like the watery seas above, the story will change form. It will likely fall to pieces, contracting for a while into something less substantial. It will probably be a story of destruction and loss, of uncertainty and extinction. But because they are awash with variability, it will also be a story of survival. The genetic underpinnings of corals are so rich, it's as if their DNA larders are stocked for the coming crisis. Not all will die. They never do. Every time the water heats, every time there is a mass bleaching, there will be those that resist. Always. Even with our powerful tools of genetics and statistics, we haven't been able to extract the secret to that story. It is still locked away deep inside the coral. As the waves continue to roll above, we are denied the satisfaction of knowing the ending.

Misha's words about the genetics of thermal tolerance being spread across the genome like butter over bread returned to me often. So many problems had come to the forefront during the pandemic. Failures of the medical system and institutionalized racism were unavoidable, along with the continued lack of response to climate change. So much of my relentless reading had shown just how interconnected these problems were. How climate change was really a problem that disproportionately affected people of color. How climate refugees will be forgotten by the medical system. How people of color died during the pandemic at almost twice the rate of white people. How as a group Hispanic, Black, and Indigenous children made up 75 percent of juvenile deaths from COVID-19, while representing only 41 percent of the population. How the pandemic was exacerbating mental illness. There's rarely just one reason, just one gene, for anything. It's so much more often system wide, genome wide. We know this in our mind, though we yearn for simple answers with our heart.

After weeks of playing Lily's Garden, I was at level 484. Lily, who had initially seemed so clever and likable, was deceived by a thin plot line. Her romance with the burly but unreliable next-door neighbor hadn't gotten off the ground. A squirrel character had been introduced. It's always a bad sign when they bring in the mammal. Remember the monkey in *Friends*? Hating myself more every time I opened the app, I scoured the internet for a way to cheat. Someone must have posted the story line so I could cure myself of the need to know what happens next. On a derelict list-serve, I found others who had succumbed to the allure of rose-adorned bricks and snail-eating bombs. And I learned that there was no ending. Even a thousand levels into the game, two thousand levels in, it never ended. In that moment, I was released. The game lost its urgency. It was no balm. Like the moment we were in, both aboveground and beneath the seas, there were no easy answers.

39

Brightening

Three and a half months after we'd left Isy, Keith and I flew to Chicago. Now that masks were available, we wore tight-fighting N95s around our faces. In my purse, I had one more, still in its plastic wrap, for Isy, for the trip home. The middle seats were empty on the flight and, like me, the airports felt abandoned and unsettled. In our rental car, we stopped at a Dunkin' Donuts to buy coffee, trying to chase away the forlorn feeling that had settled over the world with a bit of caffeine. Even the chairs and tables piled in the corners looked uncomfortable.

We checked in to the same nearly empty hotel where we'd stayed when we'd dropped Isy off. I pulled back the curtain on our fifth-floor window, revealing gray, fuzzy clouds and a parking lot. But beyond that was a hillside where yellow flowers of ragweed blossomed and cream-colored butterflies flit among the grasses. From that vantage point, I could see into the top of a streetlight, a hollow pipe whose cap had blown away. A little brown wren flew inside and then popped out; it was building a nest. That structure, I thought, was a kind of repurposing, a spontaneous restoration, a support for new life. How much of the Isy we knew before would be restored? How much would be changed?

The next day, the gray clouds had cleared, replaced with fluffy white ones perched in a fresh blue sky. As we pulled into the parking lot at the residential therapy facility, Keith said, "If we make it through the day without leaving our trunk wide open, it'll be a win."

Even though masks had become more routine over the last three and a half months, wearing one still felt unnatural. That first discordance of Isy-morphing-to-patient returned to me as I pulled the covering over my face. But then, pushing through the heavy door was a staff person holding a clipboard, and behind her was our daughter, green eyes, hair tossed in a messy bun, masked. There wasn't any question, we were hugging despite it all, despite the virus, the travel, the potential exposures. I embraced her and felt the familiar sweet urgency in her hug.

In the nearby town, we found a restaurant that was partially open and read the menus. None of us had been out to eat in more than a hundred days, and everything looked good. As we waited for our food, I found myself touching Isy's arm, her shoulder, rubbing her back. "Mom," Isy said, "you're so anxious. It's like we've changed places. When you dropped me off, I was so anxious I was always hanging on you, and you found it so annoying. Now . . ."

She let the end of the sentence hang, but I could fill in the blank. I was the one who was annoying. I could not have been more grateful to be able to be annoying.

After dinner, we dropped Isy off for her last night in the unit. We'd formally discharge her in the morning. On the way back to our hotel, Keith suggested we get a drink. We found a nearly empty hotel bar fashioned after an Irish pub with stenciled mottos high up on the walls near the ceiling encouraging friendship and fraternity in a green Celtic-styled font. The masked bartender took our orders and asked in a friendly way if we were from out of town. When we hesitated, he offered, "Visiting family?"

"Yes, our daughter." How do you tell the story casually?

The bartender brought us our drinks and left us alone. Keith lifted his scotch on the rocks. "We're getting our daughter back."

I touched the edge of my glass to his. "She seems changed, doesn't she?"

"Yes, stronger."

"That comment about me being the anxious one . . ."

We laughed and drank and managed to feel a bit celebratory. After a while, we asked for the bill. The bartender lingered while I signed.

"Think you'll come back to visit your daughter again? We've got a great hotel here." I guessed business had taken a hit from the pandemic.

"Ah, no. We're actually taking our daughter home." I just decided to come out with it. "She's been in a residential program, for OCD."

He nodded. I noticed that his eyes held a kind of sadness. "It's a world-class place."

"Yeah, seems like it. Even so, I hope we don't have to come back."

He nodded again. "My son was there, too."

"Oh," I said. "So you know about it."

He nodded. "Twice." He slid the signed receipt toward himself.

"Thank you," I said, and I understood two things. Although we needed to celebrate our victories, our fight would continue. And mental health, like coral, might be invisible, but there's nothing gained by keeping quiet about either of them. They are the foundation on which so much else rests.

THE DAY ISY CAME HOME, something else arrived in Austin: a giant dust storm. The plume originated in the Sahara Desert, the western edge of Africa, five thousand miles away. Micron-sized particles streamed across the Atlantic Ocean, over the ailing coral reefs of the Caribbean. It swept through the Gulf of Mexico and wafted into central Texas. While dust storms like this happen all the time, this one was the largest since satellites started tracking them in 2002, blowing away the previous record by 50 percent. Over the Atlantic Ocean, researchers saw changes brought by the dust scattering light back to outer space. Seawater temperatures dropped by 0.4°C. I felt a difference in Austin, and temperature records supported my hunch. The week of the storm, the average high temperature dropped by 2.5°C (4°F) compared with the week before. At night, the moon was a fuzzy approximation of its usual crisp-edged shape. During the day, the sky was bleached of all its usual robin's-egg blue. When I looked upward, I marveled at the very presence of African sands overhead, a rare visualization of the planet's winds and our interconnections. I thought about how I had missed the first demonstration of cloud-brightening technology in Australia. But I took solace in being witness to the massive-scale cloud brightening that Nature had undertaken right above my own head.

40

The Coin Toss

On September 16, 2020, there was a huge announcement about the future of coral reefs. But like so much else, it seemed to get lost in the news of the coronavirus and the U.S. presidential election. Even I, who had been following the story of coral carefully, heard about it only the next day, when Richard Vevers sent me an email. At the beginning of his message, he asked how Isy was doing.

"Things are better," I replied. "I'm not saying it's easy, and most days her OCD finds a way to ruin something. But it feels like we are on the other side of the worst of it."

School had been in session online for a few weeks. The first two days I hid on the floor of Isy's bedroom, out of view of her computer's camera, listening to teachers explain how their virtual classes would run. I empathized with the teachers who, rather than gauging students' learning by looking at the faces in a classroom, were instead looking at boxes that were mostly just names on a gray rectangle. But for Isy, the distance made rejoining academics easier. The stress of being physically in a school was removed. After a while, using the tools she'd learned in Wisconsin, she was able to do her assignments alone and turn them in. No screen grabs involved. It would never be easy and would require consistent effort, but the results were beginning to show, much like a successful restoration.

Richard wrote back, "There was a meeting about the coral reef fund yesterday that you might be interested in," with a link. Clicking led to an announcement of a massive initiative: the Global Fund for Coral Reefs, or GFCR. The goal was to fund the health of coral reefs at the greatest level ever: $500 million.

GFCR consisted of two parts: $125 million in grants and $375 million in private investment. The money would create a portfolio of businesses supporting coral restoration and conservation, which would open a pipeline to attract $2 billion to $3 billion in additional investment. The timeline was one decade. Several United Nations agencies were involved, including the UN Development Programme, the UN Environment Programme, and the UN Capital Development Fund. Private entities had signed on, including one of the biggest banks in Europe, BNP Paribas, as well as investment funds that focused on sustainability. The International Coral Reef Initiative, which included the Coral Restoration Foundation that had hosted the Reef Futures meeting in 2018, had played a critical role. The effort had been catalyzed by two major philanthropies. One was the Paul G. Allen Family Foundation, which had driven the Coral Atlas effort to map the world's coral reefs. The other was the Prince Albert II of Monaco Foundation.

Monaco, perched on the edge of the Mediterranean, has been an active player in the world of ocean science for more than a century. One of the world's great oceanographic museums, a huge white Baroque structure adorned with the names of twenty oceanographic research vessels, towers above the sea, just around the corner from the high-dollar gambling casinos of Monte Carlo. Inside, the scientific contributions of Prince Albert I, who established the museum and sailed its research vessels, are documented, alongside collections of four thousand species of fish and two hundred families of invertebrates. Jacques Cousteau was the museum's director for thirty years. Somewhere in a box of photos, I have a picture of me as a graduate student visiting the museum. I'm sitting inside one of the first deep-sea submersibles, a metal sphere with wood paneling and fanlike propellers sprouting out at odd angles.

For the announcement of the Global Fund for Coral Reefs, Prince

Albert II appeared in a room resplendent with gold-edged paneling and gilded antique furniture. He had the bearing of a business school professor, wearing wire-rimmed glasses and a dark suit and tie, citing reports and facts in a flawless American accent. "In order to have a chance to save corals, we would need to take action within the next decade. However, we know that one of the main issues encountered by coral stewards is lack of resources. Faced with such a situation, we cannot remain inactive." Prince Albert II spoke of the need to work on both the science—the search for corals that can withstand our future seas—and the deployment, which would support marine protected areas and coral restoration. However, he was clear about the underlying issue. "Above all, I'm thinking of the fight against climate change and its consequences," he said. "We cannot ignore the fact that it is by reducing the damage of a carbon economy that we will be able to protect corals sustainably."

The event continued for two hours, with leaders and officials from countries around the world committing their support. Their comments echoed one another: Coral reefs supported more people and species than any other ecosystem, yet suffered the most. Coral reefs have long been disproportionately undervalued, and we have gravely underinvested in their health. We can't imagine a world without coral. This is the moment for a seismic shift.

I wondered whether this was—in all its formality and diplomacy—the crazy idea that reefs really needed. It was the kind of mixed private-public funding model that Elinor Ostrom had espoused. It tapped into the large banks of United Nations funding that Richard Vevers had envisioned. It brought together businesses that people like Frank Mars and Gloria Fluxà had recognized as critical. The scale of the financing proposed swamped what the XPRIZE, for all its glamour, had offered. It incorporated the expertise of coral managers and scientists. What was most notable to me was that the plight of coral had reached a level of international visibility never before achieved.

As I sat at my desk, writing notes from the meeting, I heard, "Mom, can I have some help?"

"Hang on," I answered, giving myself a few seconds to finish my sentence and then put my thoughts on hold.

I walked to Isy's bedroom, where she was taking online school. "I have a practice quiz for chemistry," she said. "I'm anxious. Can you help me?"

"Sure," I answered, standing behind her so I could see her computer screen.

She clicked on the first question and read it out loud: "What is a positively charged ion?" She answered herself, "A cation." She clicked the answer and a green checkmark appeared. The screen flashed the next question. "When electrons are transferred from a metal to a nonmetal . . . a covalent bond." Click. Check.

Isy didn't need my help, just my presence behind her. As she worked, I absentmindedly ran my fingers through her hair, which had grown long during quarantine. I undid a few tangles and then pulled her hair back from her forehead and divided it into three. I gently began to pull the pieces together. When I finished, I admired the interlocking pieces, the golden and brown strands blended together. I thought about the structure of the proposed Global Fund for Coral Reefs. Philanthropy would kick it off, but it would fall apart without the support of a major funder like the United Nations and input from private industry and banks. Ideas like blue bonds, debt swaps, and reef insurance would be part of that investment. The discussion of building financial pipelines returned me to the $10,000 that Richard Vevers had struggled to invest when we visited restoration sites in Bali. I thought of how the many nations where coral reefs exist had pledged support, adding to the effort in various ways: through marine protected areas, through monitoring, mapping, and with coral restoration projects. I thought back to my enthusiasm over the XPRIZE as an inciting force to shift attention to coral. I recognized how my optimism had faded, that I no longer believed a single prize had the power to change the fate of the coral reefs. If there was to be success at scale, it would require this kind of multipronged effort. I wondered if the Global Fund for Coral Reefs would fuel a wave that gathered strength and coalesced into something meaningful. Or would it be another swell that crashed against the shore?

Holding the end of Isy's braid together with one hand, I reached up and pulled the ponytail holder out of my own hair with the other. I fastened the end of the braid and let it fall to Isy's back. Pausing from her work, Isy reached back and felt her hair. Her hand stopped on the rubber band. She pulled it around to look.

"It's a black ponytail holder," she said, her voice edged with anxiety.

"Yeah. So?" I said.

"Black ponytail holders are contaminated."

"Hmm," I said, dispassionately. I'd learned tools over the years too. Saying less was usually better. I couldn't fight Isy's OCD for her. Only she could.

When Isy's symptoms started, we had desperately wanted to find a simple answer: a pill or a treatment that would magically dissolve her problems. But in the same way that Isy had to do the hard work of checking into a residential program and learning how to restructure her response to triggers, humanity needs to reassess our fundamental behavior to solve the problem of the global coral epidemic. Like the medicines and infusions we gave Isy, coral restoration is only part of the solution. The medicines and infusions we gave Isy might have taken the edge off, so that when she got to residential treatment, she could learn the behavioral changes that made the difference. Coral restorations will be the lifeboats sheltering the coral until we get control of our carbon emissions.

In addition to the science and funding opportunities for coral, the Global Fund for Coral Reef announcement included a new, unavoidable theme: the coronavirus and its lessons. This was best articulated by the UK minister for Pacific and the Environment, who said, "This pandemic has exposed our vulnerability, economically of course, but it is also a wake-up call to profoundly reset our relationship with the natural world. Coronavirus is almost certainly the consequence of our abuse of nature. The scientific evidence is very clear that the current crisis will be dwarfed by others, by climate change and environmental degradation. Every nation has a choice for how we reboot. You can choose to lock in decades of environmental destruction, or we can turn this grim experience into something good."

Like our struggles with Isy and the insistent bleaching of the corals, the COVID-19 pandemic laid bare weaknesses. It showed problems that were already lurking beneath the surface—problems with our health care system, problems with inequality, problems with climate change. But like the way residential treatment upended the small world of our family and then brought it back together again, the devastation brought by the COVID crisis is an opportunity to rebuild better. In Europe, an economic recovery package worth $800 billion required energy-efficient building and a transition away from fossil fuels. In the United States, corporations recognized the fall in oil prices that accompanied the pandemic as a catalyst. BP, perpetrator of the largest marine oil spill in U.S. history, saw revenues fall $16.8 billion in the second quarter of 2020. So BP cut oil and gas production by 40 percent and upped its commitment to low-carbon energy technologies tenfold. The Biden administration, during its first week in office, rejoined the Paris Climate Accords and issued a slew of executive orders addressing climate change. We saw that we have power over our atmosphere. It's not too big.

In our seas, perhaps that shift had also started happening. At the end of 2020, significant efforts to support the health of our oceans were announced, again too quietly for most of us to notice amid the pandemic and the contentious presidential election. Fourteen governments responsible for 40 percent of the world's coastlines—including Australia and Indonesia, but not the United States—pledged to end illegal fishing and implement science-based fisheries plans. Separately, the United States proposed rules to add an additional six thousand square miles to marine protected areas to support the health of coral reefs.

As I stood next to her, Isy considered the ponytail holder for another couple of seconds. I saw her use one of the tools she'd practiced. She breathed deeply as she rode the wave of anxiety up and waited for it to fall.

"You know, Mom," Isy said. "When I was really sick, I lied to you all the time. I lived in a world of lies. You can't even imagine what it was like."

"I know," I said. Then we talked. She began telling the story of what it was like to live through those painful days, bringing some of it into the light. Not all of it, but a start.

Isy's mental illness and problems in the coral reefs have many similarities—being simultaneously invisible and foundational to well-being among the most obvious—but there's one glaring difference. When it comes to mental illness, we rarely have the luxury of knowing the underlying neurological cause. The same is not true for the problems with coral. We may not yet know the details of the breakup of the badass merger of the coral and the algae, but we know what's causing the loss of coral reefs: mostly climate change, but also destructive fishing, pollution, and disease. And we also know how to solve these things: stop burning fossil fuels, manage fishing, prioritize sanitation, and control pollution.

I once heard that hope and grief are two sides of the same coin—that without hope, there is no grief. And without grief, there is nothing to hope for. I believe that to be true. Grief comes from looking back at mistakes, and hope can be found only by looking forward. When I think of corals, and of our planet, I know that I do not yet feel the full weight of grief. There is still too much left to hope for. But I do feel as if the coin has been tossed in the air—or rather many coins have been tossed in the air for the many reefs around the world. I can see them whirling and spinning up there, and I know that they cannot escape the pull of gravity. The coins will inevitably fall. One side will land upward. It is too early to call which side it will be.

While I was able to imagine very little about what would happen in the two years since I attended the Reef Futures meeting, I always knew that the lessons from the corals would be troubling: stories of death and loss, of neglect, of systemic problems. As I traveled and spoke with experts to explore those stories, I saw all of that: the relentless march of SCTLD into the Caribbean, the loss of reef to explosions, the destruction of coral wrought by superheated hurricanes, and the most extensive mass bleaching on record. I was also hoping to discover tools to combat those problems. And that happened too. I saw the potential for stabilization with rebar stars, rebuilding flourishing reefs. I witnessed coral grown in nurseries, and coral larvae finding homes on substrates designed to give them a head start. I watched as coral that had never spawned in captivity burst forth with new life. I found communities of people who continue to innovate around the future of coral: building frozen banks of coral genes and adding glitter to the sky to cool the reef. And I was heartened to discover that evolution may have already equipped corals, and their powerful algal allies, with tools to survive in a fevered sea. I learned about new financial tools and business ideas that might give coral a chance in increasingly hostile waters, and the beginnings of a global fund to support the future of coral reefs.

But when I look back, the most powerful lesson is what's at the very root of the coral's badassness: the lesson of cooperation. If this book had been the story of an XPRIZE winner, it would have ended with a single strand, a single team claiming victory. Only by braiding together the many tools we have and only by using whatever language it is that we share—be it the nitrogen and sugar of the coral and algae or the many words that pass between us as humans—can we have hope of adding weight to any of the spinning coins, nudging them toward survival as they inevitably fall.

Isy let go of the black rubber band she'd been holding. She tossed the braid down her back and turned to her schoolwork. "An ion made of two or more atoms is a . . . polyatomic ion." Click. Check.

She looked over at me and gave a soft smile. "You can go."

AS COVID VACCINES BECAME AVAILABLE IN THE SPRING, I emerged hesitantly from isolation at first and then with growing relief and eagerness. Travel out of the country was still restricted, but there was one reef right in my own backyard that had, like so many coral stories, made big news quietly. On his second-to-last day in office, President Trump brought to a close an effort of more than fourteen years, one that had spanned two

Republican and two Democratic presidencies. With his signature, he expanded protections for the Flower Garden Banks National Marine Sanctuary, located one hundred miles offshore from Galveston in the Gulf of Mexico, tripling its size. One of the few ways to see the coral there is by way of the single dive boat that makes regular trips to the Sanctuary. I signed up our family for an early May trip.

The twenty-six species of coral in the Flower Garden Banks grow on tops of undersea mountains of salt, called salt domes, that rise from depths of three or four hundred feet to within sixty feet of the sea surface, where there's enough light to fuel the badass merger. That salt is both the remains of evaporated ancient oceans and also the vault that sealed in place the rich oil deposits that made Texas a fossil fuel powerhouse. Ever since the reefs of the Flower Garden Banks were first explored in the 1960s, their fate has been connected to the oil and gas industry. As drilling platforms marched farther and farther offshore, the government restricted operations near the coral and even required that oil and gas companies fund research into this unique Caribbean ecosystem. The recent expansion of those protections was spearheaded by a staunch Republican from Houston, a petroleum engineer and businessman named Clint Moore. Moore also had a deep passion for the Flower Garden Banks, especially for the spectacular manta rays that depend on the coral reefs as juvenile nursery grounds.

As the western U.S. suffered through a brutal heat wave linked to climate change, in Texas an unseasonably stormy spring delayed our trip. When we were able to reschedule, it had been almost a year since Isy's discharge from the mental health facility in Wisconsin. But she wouldn't be coming with us. She had already left to be a counselor at an overnight summer camp. She was separated from our family again, but this time building a life for herself.

Keith, Ben, and I drove down to the coast and boarded the hundred-foot dive boat at sunset. Our itinerary called for traveling overnight to reach the Sanctuary. As the boat motored down a humid, mosquito-buzzing canal, industrial shapes of refineries lit up around us. In the cabin's bunkbed, I struggled to fall sleep, as much from the strange sounds of roaring engines and slapping waves as from my anticipation at seeing the reefs, which had taken on a certain mystique. Ten thousand years ago, Caribbean coral at least four hundred miles away spawned larvae that had surfed the currents and landed, almost inconceivably, on the tops of the Gulf of Mexico's salt domes, forming the basis for a teeming ecosystem. Since the late 1980s, monitoring surveys of the Flower Garden Banks had identified more than 280 species of fish, 60 species of crustaceans, and more than 600 species of mollusks. At least two species of manta rays, eagle rays, loggerhead turtles, hawksbill turtles, hammerheads, and whalesharks were frequent visitors. Most remarkably, as opposed to nearly everywhere else in the world, the studies had shown an *increase* in coral cover, from 50 to 60 percent. The salt domes' distance from shore and depth had seemed to protect the corals.

But it's possible that isolation might not be enough for much longer. In 2016, part of the Flower Garden Banks was smothered by low-oxygen water thought to have origins in heavy spring-storm runoff, indicating the reef might not be so safe from shore-borne impacts after all. Flower Garden Banks coral has managed to avoid severe bleaching, although there have been occasional instances of bleaching, and seawater temperatures have increased about a degree. So far, the coral has also managed to avoid the diseases that have culled so many colonies in the Caribbean. SCTLD hasn't yet arrived, but it lurks nearby.

In the morning, I awoke as we passed through a small squall. Walking out onto the back of the boat, I noticed a rainbow curving through the dawn sky just as it had so many years ago on the Caribbean cruise with my grandmother. I thought of its rarely seen circular form, and how this story had returned me to the too frequently unseen coral reefs that I love but also given me perspective on their plight. We moored at the West Flower Garden Bank, one of the two coral-laden salt domes protected by the Sanctuary. The dive officers instructed us to follow a yellow rope down to the reef and across the coral, like following a path through the woods. We descended into crystal-clear blue water as warm as a bath. Sixty feet down, fields rich in coral stretched as far as I could see. Massive plates covered the seafloor, so continuous that they bumped up against one another, the spaces in between colonies marked by their stinging-cell-fought territorial wars. I soared over the reef, astonished at the extent of the coral. I spread out like a sea star, and the corals were bigger than my wingspan. Colonies this size could be a thousand years old, maybe more. When I looked up, the sea was teeming with pastel parrotfish, winged triggerfish, and the silhouettes of jacks and chubs. When I looked down, it was bobbing with hermit crabs, winking with Christmas tree worms, and crawling with candy-cane-banded shrimp.

In the afternoon, our dive boat moved to the East Flower Garden Bank. Again I descended down the yellow rope. Here, the coral took the forms of car-sized hills, each one draped in enormous coral colonies that frilled like lacy skirts at the base. This place wasn't so much for soaring as for careening up and down and over. It was for finding bright-eyed squirrelfish and yawning moray eels. It was for marveling at the long quivering spines of sea urchins. It was for watching the bobbling of a spotted trunkfish, a creature that seems like it shouldn't be able to swim at all.

I came to a halt before a looming hill of coral and forced my flippered feet downward as if I were standing. The coral rose majestically before me, and I gazed up to take in its full height. Then I slowly circled. I was in a grove of these giants. I felt the same awe I'd felt in the presence of redwoods, surrounded by ancients. These corals, I thought as I turned, are the matriarchs and patriarchs of the sea. That they still exist felt more miraculous than treasure at the end of a rainbow. How long they'll remain depends on us.

Acknowledgments

Like a coral reef, a book goes by one name, but it's really a compendium of interlocking, interdependent contributions. I give my heartfelt thanks to everyone who helped this book come to life.

First among those to thank are the scientists and coral experts who gave so willingly of their time and experience. Special gratitude to Richard Vevers, Stephanie Tate, Lizzie Mcleod, Noel Janetski, Alicia McArdle, Saipul Rapi, Frank Mars, Vincent Chalias, Megan Morikawa, and Macarena Blanco, who made my trips possible. To Dan Harrison and Line Bay, I hope I get there soon.

I extend my heartfelt thanks to all the scientists and experts I have mentioned in these pages. Other coral experts I wasn't able to include, but whom I'd like to thank for generously helping me understand the sea's greatest ecosystem are: Rachel Wright, Tali Vardi, Phil Dustin, George Stanley, Peggy Winkler, Nikki Traylor-Knowles, Joe Pollock, Margaret Miller, Cudo Prasetya, Camilo Cortés-Useche, Thomas Krueger, Anupa Asokan, Jonathan Landrey, Rowan Martindale, Joanna Smith, Lisa Boucher, Nancy Knowlton, Emily Jeffers, Fiona Merida, Brent Chatterton, James Guest, Malina Fagan, and Charles Sheppard. A special thank you to Deanna Soper: we need to go to another coral swap soon or, better yet, diving.

The Matz lab opened their doors—and Zoom links and backyards—and allowed me to immerse myself in the world of coral even when none of us could visit a reef. I am forever grateful to Misha Matz, J. P. Rippe, Kristina Black, Evelyn Abbott, Hayley Bedwell-Ivers, Coral Loockerman, Carly Scott, Christopher Peterson, Eunice Wong, Yi-Jyun Luo, Erik Iverson, Ali Thabet, Sofia Guajardo Beskid, and Dominique Gallery for getting me through that time. Thank you also to Susan Hovorka and The University of Texas Libraries for making scientific research available to me. If I have forgotten anyone who helped me with coral science, please know it is inadvertent, and I am grateful for all you do for the seas.

Heather Kuhlkin deserves special thanks for graciously loaning me dive gear and always being there to discuss the great underwater world. Julia Clarke, you are the very best sounding board and urban hiking partner. Thank you to the Gorelick family for sharing your beautiful aquaria with me.

I feel lucky to have met Alan Powderham at just the perfect moment. Your photos are extraordinary, and I am honored to share these pages with them. Thank you to The Ocean Agency and Sergio Guendulain-García for generously sharing your photographs as well.

My writing community has been a bedrock during a turbulent time. The Writers of the Shed, your constant feedback and encouragement made all of this possible. Thank you from the depths of my heart to Esther Mizrachi and Brittani Sonnenberg, and much love to the memory of Kebana Frost. Other writing support rolled in at key times from Thomas Hayden, Sy Montgomery, Tamar Stelling, and the members of ATXSciWri and LLL.

Fact checking is more important now than ever. I celebrate the day I asked Glynnis Collins to help me fact-check this book. You are eagle-eyed and amazing. If there are any remaining mistakes, I hope readers will let me know so they can be corrected.

It is a joy and honor to be published by Riverhead Books. Courtney Young zeroes in on everything that matters and makes all that she reads even stronger. Thanks to Jacqueline Shost for keen eyes and enthusiasm. The publicity team, especially Glory Ann Plata, got this book into the hands of many more than I could have imagined. And much gratitude to Mollie Glick, for continuing to see my vision of what science writing can be.

I am grateful to dear friends who supported me with the generosity of spaces to escape to and write during a year of confinement. Thanks to the Kings, Ferdmans, Pennells, and Handel-Hirshes.

A special thank you to the experts in mental health whose names I will omit for confidentiality. Your work is a gift, and you have taught me more than you can imagine.

This book, like life, isn't only about science, it's about family, and I am beyond grateful to be part of a family that has always been there to support me. Thank you for a lifetime of love, Gail and David Berwald, Bara Robin-Fern and Paul Fern, the Steins, Berwalds, and Ferns. Most important, I give my heart to Keith for always making me laugh, Ben for his steadfastness, and Isy for being the sweetest and toughest of all.

Notes

Chapter 1. Fairy Land of Fact

3 **"fairy land of fact"**: William Saville-Kent, preface to *The Great Barrier Reef of Australia: Its Products and Potentialities* (London: W. H. Allen, 1908).

7 **on average 0.8°C (1.4°F) warmer:** "Climate at a Glance: Global Time Series," NOAA National Centers for Environmental Information, 2021, https://www.ncdc.noaa.gov/cag/global/time-series/globe/ocean/ann/7/1880-2020.

7 **Half the Great Barrier Reef's corals:** Andreas Dietzel et al., "Long-Term Shifts in the Colony Size Structure of Coral Populations along the Great Barrier Reef," *Proceedings of the Royal Society B* 287, no. 1936 (2020), https://royalsocietypublishing.org/doi/10.1098/rspb.2020.1432.

7 **another degree warmer could be fatal:** Matthew Collins et al., "Chapter 6: Extremes, Abrupt Changes and Managing Risks," *IPCC Special Report on the Ocean and Cryosphere in a Changing Climate*, ed. H.-O. Pörtner et al., in press, 2019, https://www.ipcc.ch/srocc/chapter/chapter-6/.

7 **may not exist by 2050:** "Based on findings from simulation modelling, SR15 concluded that 'coral reefs are projected to decline by a further 70–90% at 1.5°C (very high confidence) with larger losses (>99%) at 2°C (very high confidence).'" N. L. Bindoff et al., "Changing Ocean, Marine Ecosystems, and Dependent Communities," *IPCC Special Report on the Ocean and Cryosphere in a Changing Climate*, https://www.ipcc.ch/srocc/chapter/chapter-5/.

7 **a fourth of all marine species:** Francis Staub, "How the World Is Coming Together to Save Coral Reefs," World Economic Forum, December 4, 2020, https://www.weforum.org/agenda/2020/12/how-the-world-is-coming-together-to-save-coral-reefs/.

7 **at between $2.7:** Staub, "How the World Is Coming Together to Save Coral Reefs."

7 **$10 trillion:** Twenty-eight million hectares of coral reef at $352,249 per hectare per year = 9.86 trillion. Robert Costanza et al., "Changes in the Global Value of Ecosystem Services," *Global Environmental Change* 26 (2014): 152–58, https://doi.org/10.1016/j.gloenvcha.2014.04.002.

7 **97 percent of wave energy:** Filippo Ferrario et al., "The Effectiveness of Coral Reefs for Coastal Hazard Risk Reduction and Adaptation, *Nature Communications* 5, no. 3794 (2014), https://doi.org/10.1038/ncomms4794.

7 **$1.8 billion annually:** Curt D. Storlazzi et al., *Rigorously Valuing the Role of U.S. Coral Reefs in Coastal Hazard Risk Reduction*, U.S. Geological Survey Open-File Report 2019–1027, 2019, https://doi.org/10.3133/ofr20191027. "Coral Reef Barriers Provide Flood Protection for More Than 18,000 People and $1.8 Billion Worth of Coastal Infrastructure and Economic Activity Annually," U.S. Geological Survey, 2019, https://www.usgs.gov/news/coral-reef-barriers-provide-flood-protection-more-18000-people-and-18-billion-worth-coastal.

7 **$9 billion:** Staub, "How the World Is Coming Together to Save Coral Reefs."

9 **called Reef Futures:** "Reef Futures 2018: A Coral Restoration and Intervention-Science Symposium," Coral Restoration Consortium, 2018, https://reeffutures2018.dryfta.com/.

9 **Is it worth the cost?:** Alasdair J. Edwards and Susan Clark, "Coral Transplantation: A Useful Management Tool or Misguided Meddling?" *Marine Pollution Bulletin* 37, no. 8–12 (December 1999): 474–87, https://www.sciencedirect.com/science/article/pii/S0025326X99001459.

9 **lab where I did my PhD:** "Dale Kiefer," University of Southern California Dornsife College of Letters, Arts and Sciences, https://dornsife.usc.edu/cf/faculty-and-staff/faculty.cfm?pid=1003407.

9 **registration button:** I offset carbon emissions for research travel for this book by destroying an equivalent amount of chlorofluorocarbons through Tradewater.us. CFCs are refrigerants that are greenhouse gases about ten thousand times more potent than carbon dioxide. They were banned in 1989 with the Montreal Protocol, but no mechanism for destroying existing stores was included in the treaty, and it's unclear how many still remain in rusting canisters around the world. The voluntary CFC offset market has emerged as the only way to destroy these gases before they leak into the atmosphere. Other advantages of CFC offsets are that their destruction is verifiable and permanent.

Chapter 2. Crazy Ideas

10 **Coral Restoration Consortium:** Coral Restoration Foundation, https://www.coralrestoration.org/coral-restoration-consortium. Reef Resilience Network, http://crc.reefresilience.org/.

11 **half of the Great Barrier Reef:** Andreas Dietzel et al., "Long-Term Shifts in the Colony Size Structure of Coral Populations along the Great Barrier Reef," *Proceedings of the Royal Society B,* 287 (2020): 20201432, https://royalsocietypublishing.org/doi/10.1098/rspb.2020.1432.

13 **Charles Lindbergh landed:** "Raymond Orteig–$25,000 Prize," Charles Lindbergh: An American Aviator, 2014, http://charleslindbergh.com/plane/orteig.asp.

13 **Peter Diamandis:** Raya Bidshahri, "5 Organizations Using Cool Tech Solutions and Research to Clean Up the Oceans," *Singularity Hub,* May 20, 2018, https://singularityhub.com/2018/05/20/5-organizations-using-cool-tech-solutions-and-research-to-clean-up-the-oceans/.

14 **Saving Coral Reefs:** Visioneering 2018, XPRIZE, October 22, 2018, https://www.xprize.org/articles/visioneering-2018.

15 **Ruth Gates:** Ed Yong, "The Fight for Corals Loses Its Great Champion," *The Atlantic,* October 29, 2018, https://www.theatlantic.com/science/archive/2018/10/optimist-who-believed-saving-corals/574240/. Gates Coral Lab, http://www.gatescorallab.com/.

Chapter 3. The Issue of Scale

17 **RRAP:** Australian Reef Restoration and Adaptation Program, https://www.csiro.au/en/research/animals/marine-life/Reef-restoration-program.

17 **Great Barrier Reef Marine Park Authority:** Australian Government Great Barrier Reef Marine Park Authority, https://www.gbrmpa.gov.au.

17 **"no stone unturned" approach:** A modeling study a few years later concluded that the most promising mitigation tools for the Great Barrier Reef are cloud brightening, enhancing coral thermal tolerance, and expanded control of the Crown-of-thorns starfish. Scott Condie et al., "Large-Scale Interventions May Delay Decline of the Great Barrier Reef," *Royal Society Open Science* 8, no. 4 (2021): 201296, https://royalsocietypublishing.org/doi/10.1098/rsos.201296.

21 **Florida Reef Tract:** "Florida," Coral Reef Information System, March 22, 2021, https://www.coris.noaa.gov/portals/florida.html.

22 **consortium of Florida agencies:** "Stony Coral Tissue Loss Disease Response," Florida Department of Environmental Protection, April 1, 2021, https://floridadep.gov/rcp/coral/content/stony-coral-tissue-loss-disease-response.

Chapter 4. Buying Excitement

24 **XL Catlin:** The company is now known as AXA XL.

25 **underwater street view:** The Ocean Agency, https://theoceanagency.org/projects/underwater-street-view. XL Catlin Seaview Survey, https://www.catlinseaviewsurvey.com/. Bryan Walsh, "Breaking the Waves: Catlin Seaview Survey Digitizes the Endangered Oceans," *Time,* July 31, 2013, https://science.time.com/2013/07/31/breaking-the-waves-catlin-seaview-survey-digitizes-the-endangered-oceans/.

26 **"wicked problem":** Horst W. J. Rittel and Melvin M. Webber, "Dilemmas in a General Theory of Planning," *Policy Sciences* 4 (1973): 155–69, https://archive.epa.gov/reg3esd1/data/web/pdf/rittel%2bwebber%2bdilemmas%2bgeneral_theory_of_planning.pdf.

Chapter 5. A Badass Merger

35 **"cannot make them go extinct":** Mikhail Matz et al., "Estimating the Potential for Coral Adaptation to Global Warming across the Indo-West Pacific," *Global Change Biology* 26, no. 6 (2020): 3473–81, https://doi.org/10.1111/gcb.15060.

35 **"no trade-offs":** Rachel M. Wright et al., "Positive Genetic Associations among Fitness Traits Support Evolvability of a Reef-Building Coral under Multiple Stressors," *Global Change Biology* 25, no. 10 (2019): 3294–304, https://doi.org/10.1111/gcb.14764.

35 **"identify reefs":** Mika McKinnon, "Behind the Scenes of the First Crewed American Spaceflight," Gizmodo, May 5, 2014, https://gizmodo.com/behind-the-scenes-of-the-first-manned -american-spacefli-1571908246.

36 **nutritional deserts:** "Sea Surface Temperature & Chlorophyll," Earth Observatory, NASA, January 2021, https://earthobservatory.nasa.gov/global-maps/MYD28M/MY1DMM_CHLORA.

37 **a baby coral:** Some mother corals, like those in the Hawaiian species *Montipora capitata*, do package zooxanthellae into their egg packets like lunch boxes for their offspring as they set off to find their own homes. J. L. Padilla-Gamiño et al., "From Parent to Gamete: Vertical Transmission of *Symbiodinium* (Dinophyceae) ITS2 Sequence Assemblages in the Reef Building Coral *Montipora capitata*," *PLoS ONE* 7, no. 6 (2012): e38440, https://doi.org/10.1371/journal.pone.0038440.

37 **do some advertising:** W. K. Fitt, "The Role of Chemosensory Behavior of *Symbiodinium microadriaticum*, Intermediate Hosts, and Host Behavior in the Infection of Coelenterates and Molluscs with Zooxanthellae," *Marine Biology* 81 (1984): 9–17, https://link.springer.com/article/10.1007 /BF00397620.

38 **ideal trading partners:** Thomas Krueger et al., "Temperature and Feeding Induce Tissue Level Changes in Autotrophic and Heterotrophic Nutrient Allocation in the Coral Symbiosis—A NanoSIMS Study," *Scientific Reports* 8 (2018): 12710, https://doi.org/10.1038/s41598-018-31094-1. Paul G. Falkowski et al., "Light and the Bioenergetics of a Symbiotic Coral," *BioScience* 34, no. 11 (1984): 705–9, https://doi.org/10.2307/1309663.

38 **intimacy:** One example of this tight relationship was teased out by Ruth Gates, who showed that coral release special amino acids that are like chemical request slips for algal sugar. Ruth Gates et al., "Free Amino Acids Exhibit Anthozoan 'Host Factor' Activity: They Induce the Release of Photosynthate from Symbiotic Dinoflagellates in Vitro," *Proceedings of the National Academy of Sciences* 92, no. 16 (1995): 7430–34, https://www.pnas.org/content/92/16/7430.short. Transfers of nutrients between the coral and algae can happen in as little as fifteen minutes. Christophe Kopp et al., "Subcellular Investigation of Photosynthesis-Driven Carbon Assimilation in the Symbiotic Reef Coral *Pocillopora damicornis*," *mBio* 6, no. 1 (2015): e02299-14, https://mbio.asm.org/content/6/1/e02299-14.short.

39 **In tropical oceans:** "An active reef may produce 20 grams of organic matter per square meter of surface per day. This is about equal to the maximum productivity of a very efficient terrestrial crop, such as sugar cane, growing under optimum conditions. A wheat field, or an assemblage of living organisms in a temperate marine environment, generates about one-tenth of the dry weight of organic material produced each day by a coral reef of equal area. The productivity of marine plankton in open water in the tropics is no better than that of organisms in a desert—about 0.2 gram per square meter per day." Eugene Kozloff, *Invertebrates* (Orlando, FL: Holt Rinehart and Winston, 1990), 138.

39 **Roberto Iglesias-Prieto:** Liz Allen, "Bleached Corals Compensate for Stress by Eating More Plankton," *Forbes*, June 30, 2019, https://www.forbes.com/sites/allenelizabeth/2019/06 /30/bleached-corals-compensate-for-stress-by-eating-more-plankton/#447a429321ac.

Chapter 6. Hopeful Monsters

41 **picture book of the continents:** Alain Manesson Mallet, *Description de l'Univers* (Paris, 1683), via Professor Emerita Frances W. Pritchett, Columbia University, http://www.columbia.edu/itc /mealac/pritchett/00generallinks/mallet/index.html#index.

41 **stint as a schoolteacher:** Alain Manesson Mallet, Wikipedia, accessed March 4, 2021, https:// en.wikipedia.org/wiki/Alain_Manesson_Mallet#cite_note-2.

42 *Corals of the World:* J. E. N. Veron, Mary Stafford-Smith, ed., *Corals of the World* (Townsville: Australian Institute of Marine Science, 2000), 3 vols. The book is available online and the taxonomy discussion referenced is J. E. N. Veron, M. G. Stafford-Smith, E. Turak, and L. M. DeVantier, Corals of the World: Overview of Coral Taxonomy, 2016, Version 0.01 (Beta), accessed July 6, 2021, http://www.coralsoftheworld.org/page/overview-of-coral-taxonomy/?version=0.01.

42 **Charlie Veron:** Tim Elliott, "Live near the Beach? Coral Reef Expert Charlie Veron Has Some Advice for You," *The Sydney Morning Herald*, July 14, 2017, https://www.smh.com.au/lifestyle/charlie-veron-the-dire-environmental-prognosis-we-cannot-ignore-20170711-gx8tqr.html. Charlie Veron, *A Life Underwater* (Melbourne: Penguin Random House Australia, 2017).

43 **Jean-André Peyssonnel:** Jean-André Peyssonnel, *Traduction d'un article des transactions philosophiques sur le corail: Diverses Observations Sur les courans de la Mer, faites en différens endroits* [Translation of an article on the philosophical transactions of coral: Various Observations on the Currents of the Sea, Made in Different Locations] (London: Royal Society of Sciences, 1756), Eighteenth Century Collections Online, accessed July 7, 2021, https://www.gale.com/primary-sources/eighteenth-century-collections-online.

45 **what Darwin had in mind:** David Dobbs, *Reef Madness: Charles Darwin, Alexander Agassiz, and the Meaning of Coral* (New York: Pantheon Books, 2005), 107.

45 **these "hopeful monsters":** Bridget Alex, "How Do We Define a Species?" *Discover*, March 19, 2019, accessed March 4, 2021, https://www.discovermagazine.com/the-sciences/how-do-we-define-a-species.

45 **already risen by a degree:** M. A. Alexander et al., "Projected Sea Surface Temperatures over the 21st Century: Changes in the Mean, Variability and Extremes for Large Marine Ecosystem Regions of Northern Oceans," *Elementa: Science of the Anthropocene* 6 (2018): 9, https://www.elementascience.org/articles/10.1525/elementa.191/.

45 **rise by three feet:** Rebecca Lindsey, "Climate Change: Global Sea Level," NOAA Climate.gov, January 25, 2021, https://www.climate.gov/news-features/understanding-climate/climate-change-global-sea-level.

45 **too conservative:** University of Copenhagen, "Sea Level Likely to Rise Faster Than Previously Thought," Phys.org, February 2, 2021, https://phys.org/news/2021-02-sea-faster-previously-thought.html.

Chapter 7. Tiny Architects

47 **Darwin figured:** Frederick Burkhardt, "Darwin & Coral Reefs," Darwin Correspondence Project, University of Cambridge, 2020, https://www.darwinproject.ac.uk/commentary/geology/darwin-coral-reefs.

48 **karstification:** André W. Droxler and Stéphan J. Jorry, "The Origin of Modern Atolls: Challenging Darwin's Deeply Ingrained Theory," *Annual Review of Marine Science* 13, no. 21 (2021): 1–37, https://doi.org/10.1146/annurev-marine-122414-034137.

51 **twenty-two million tons each day:** The Ocean Portal Team, Reviewed by Jennifer Bennett (NOAA), "Ocean Acidification," *Smithsonian*, April 2018, https://ocean.si.edu/ocean-life/invertebrates/ocean-acidification.

51 **that build calcium carbonate structures:** Andrew Alden, "Calcite vs Aragonite," ThoughtCo, August 27, 2020, https://www.thoughtco.com/calcite-vs-aragonite-1440962.

51 **use extra energy:** Peter Hannam, "'Death Blow': Corals, Algae Don't Acclimatise to More Acidic Seas," *The Sydney Morning Herald*, May 28, 2019, https://www.smh.com.au/environment/climate-change/death-blow-corals-algae-don-t-acclimatise-to-more-acidic-seas-20190527-p51rmn.html.

51 **400 parts per million:** Andrew Friedman, "The Last Time CO_2 Was This High, Humans Didn't Exist," Climate Central, May 3, 2013, https://www.climatecentral.org/news/the-last-time-co2-was-this-high-humans-didnt-exist-15938.

51 **coral at lower pHs:** O. Hoegh-Guldberg et al., "Coral Reefs under Rapid Climate Change and Ocean Acidification," *Science* 318, no. 1737 (2007), https://fish.gov.au/Archived-Reports/2014/Documents/Hoegh_guldberg_et_al_2007_coral_reef_climate_change.pdf. F. Marubini et al., "Dependence of Calcification on Light and Carbonate Ion Concentration for the Hermatypic Coral *Porites compressa*," *Marine Ecology Progress Series* 220 (2001): 153–62, https://www.researchgate.net/publication/250217473_Dependence_of_calcification_on_light_and_carbonate_ion_concentration_for_the_hermatypic_coral_Porites_compressa. C. B. Bove et al., "Common Caribbean Corals Exhibit Highly Variable Responses to Future Acidification and Warming," *Proceedings of the Royal Society B: Biological Sciences* 286, no. 1900 (2019): 20182840, https://royalsocietypublishing.org/doi/pdf/10.1098/rspb.2018.2840. Nathaniel R. Mollica et al., "Ocean Acidification Affects Coral Growth by Reducing Skeletal Density," *Proceedings of the National Academy*

of Sciences of the United States of America 115, no. 8 (2018): 1754–59, https://www.pnas.org/content/115/8/1754. Ana Martinez et al., "Species-Specific Calcification Response of Caribbean Corals after 2-Year Transplantation to a Low Aragonite Saturation Submarine Spring," *Proceedings of the Royal Society B: Biological Sciences* 286, no. 1905 (2019): 20190572, https://royalsocietypublishing.org/doi/full/10.1098/rspb.2019.0572.

51 **system of pH control:** Laetitia Plaisance, "Sneak Peek: Future of Coral Reefs in an Acidifying Ocean," *Smithsonian*, August 2012, https://ocean.si.edu/ecosystems/coral-reefs/sneak-peek-future-coral-reefs-acidifying-ocean. Malcolm McCulloch et al., "Coral Resilience to Ocean Acidification and Global Warming through pH Up-Regulation," *Nature Climate Change* 2, no. 8 (2012): 623–27, https://www.researchgate.net/publication/229429812_Coral_resilience_to_ocean_acidification_and_global_warming_through_pH_up-regulation.

52 **biological chisels:** J. Stanley Gardiner, "Photosynthesis and Solution in Formation of Coral Reefs," *Nature* 127 (1931): 857–58, https://doi.org/10.1038/127857a0. Jerimiah Oetting, "Dead Reefs Keep Calcifying but Only by Day," *Eos*, December 24, 2019, https://eos.org/articles/dead-reefs-keep-calcifying-but-only-by-day.

52 **without living coral eroded:** Smithsonian Tropical Research Institute, "Living Coral Cover Will Slow Future Reef Dissolution," *ScienceDaily*, September 26, 2019, https://www.sciencedaily.com/releases/2019/09/190926141713.htm.

52 **In Florida:** Nancy Muehllehner et al., "Dynamics of Carbonate Chemistry, Production and Calcification of the Florida Reef Tract (2009–2010): Evidence for Seasonal Dissolution," *Global Biogeochemical Cycles* 30, no. 5 (2016): 661–68, https://agupubs.onlinelibrary.wiley.com/doi/full/10.1002/2015GB005327.

52 **in the Pacific:** Chelsea Harvey, "Corals Are Dissolving Away," E&E News, *Scientific American*, February 23, 2018, https://www.scientificamerican.com/article/corals-are-dissolving-away1/. Bradley D. Eyre et al., "Coral Reefs Will Transition to Net Dissolving before End of Century," *Science* 359, no. 6378 (2018): 908–11, https://science.sciencemag.org/content/359/6378/908.

52 **Darwin's fight:** David Dobbs, *Reef Madness: Charles Darwin, Alexander Agassiz, and the Meaning of Coral* (New York: Pantheon Books, 2005), 146.

52 **By 2100:** Christopher E. Cornwall et al, "Global Declines in Coral Reef Calcium Carbonate Production under Ocean Acidification and Warming," *Proceedings of the National Academy of Sciences of the United States of America* 118, no. 21 (2021): e2015265118, https://www.pnas.org/content/118/21/e2015265118.

Chapter 8. Bleaching Beginnings

55 **Alfred Goldsboro Mayor:** The spelling of his middle and last names varies. They are written as Goldsborough and Goldsboro, and Mayor and Mayer. Charles B. Davenport, *Biographical Memoir Alfred Goldsborough Mayor 1868–1922*, Memoirs of the National Academy of Sciences, vol. 21, eighth memoir (Washington, DC: United States Government Printing Office, 1927), http://www.nasonline.org/publications/biographical-memoirs/memoir-pdfs/mayor-a-g.pdf. Photos of Loggerhead Key and the memorial to Mayor are found in Eugene A. Shinn and Walter C. Jaap, *Field Guide to the Major Organisms and Processes Building Reefs and Islands of the Dry Tortugas: The Carnegie Dry Tortugas Laboratory Centennial Celebration (1905–2005)*, October 13–15, 2005, https://pubs.usgs.gov/of/2005/1357/report.pdf.

56 **first rigorous experiments:** Alfred G. Mayer, "Toxic Effects Due to High Temperature," Carnegie Institution of Washington Publication 252 (1918): 173–78, https://www.biodiversitylibrary.org/item/17777#page/239/mode/1up. Alfred G. Mayer, "The Effects of Temperature upon Tropical Marine Animals," Carnegie Institution of Washington Publication 183 (1914): 5, 6, 20, 21, https://www.biodiversitylibrary.org/item/53534#page/19/mode/1up.

56 **93 percent of the Earth's warming:** Kendra Pierre-Louis, "2019 Was a Record Year for Ocean Temperatures, Data Show," *The New York Times*, January 13, 2020, https://www.nytimes.com/2020/01/13/climate/ocean-temperatures-climate-change.html. Lijing Cheng et al., "Record-Setting Ocean Warmth Continued in 2019," *Advances in Atmospheric Sciences* 37 (2020): 137–42, https://link.springer.com/article/10.1007%2Fs00376-020-9283-7. LuAnn Dahlman and Rebecca Lindsey, "Climate Change: Ocean Heat Content," NOAA Climate.gov, August 17, 2020, https://www.climate.gov/news-features/understanding-climate/climate-change-ocean-heat-content.

56 **more than three nuclear explosions:** Ivana Kottasová, "Oceans Are Warming at the Same Rate as If Five Hiroshima Bombs Were Dropped in Every Second," CNN, January 13, 2020, https://www.cnn.com/2020/01/13/world/climate-change-oceans-heat-intl/index.html. Cheng, "Record-Setting Ocean Warmth Continued in 2019."

56 **Vaughan's expertise:** Thomas Gordon Thompson, "Thomas Wayland Vaughan," *Biographical Memoirs*, vol. 32 (New York: Columbia University Press for the National Academy of Sciences of the United States of America, 1958), 399, http://www.archive.org/stream/biographicalmemo 012026mbp/biographicalmemo012026mbp_djvu.txt.

56 **corals first turned pale:** T. W. Vaughan, "Reef Corals of the Bahamas and of Southern Florida," *Carnegie Institution of Washington Year Book No. 13* (Washington, DC: Gibson Brothers, 1914), 222–26, https://archive.org/details/yearbookcarnegie13carn/page/222/mode/2up.

57 **In a picture:** B. Morton, "Charles Maurice Yonge: 9 December 1899–17 March 1986," *Biographical Memoirs of Fellows of the Royal Society* 38 (1992): 377–412, https://royalsocietypublishing .org/doi/10.1098/rsbm.1992.0020.

57 **one of those low tides:** C. M. Yonge and A. G. Nichols, "Studies on the Physiology of Corals: IV. The Structure, Distribution and Physiology of the Zooxanthellae," *Great Barrier Reef Expedition 1928–1929, Scientific Reports*, vol. 1 (London: British Museum of Natural History, 1931), 154–55, 166, https://www.biodiversitylibrary.org/page/42820138#page/225/mode/1up.

59 **"There is thus evidence":** Yonge and Nichols, "Studies on the Physiology of Corals," 157.

59 **offered Maurice a job:** Morton, "Charles Maurice Yonge: 9 December 1899–17 March 1986," 392.

59 **"who will walk":** Morton, "Charles Maurice Yonge: 9 December 1899–17 March 1986," 396.

59 **traveled to Low Isles:** Maoz Fine et al., "Ecological Changes over 90 Years at Low Isles on the Great Barrier Reef," *Nature Communications* 10, no. 4409 (2019), https://www.nature.com/articles /s41467-019-12431-y. Bar-Ilan University, "Longest Coral Reef Survey to Date Reveals Major Changes in Australia's Great Barrier Reef," *American Institute of Physics*, September 27, 2019, https://phys.org/news/2019-09-longest-coral-reef-survey-date.html.

Chapter 9. Bleaching Bombardment

61 *zonation:* For example, Eugene P. Odum, *Fundamentals of Ecology* (Philadelphia: Saunders College Publishing, 1971), 330–48.

61 **The Hawaiian Electric Company:** "Our History and Timeline," Hawaiian Electric, 2021, https://www.hawaiianelectric.com/about-us/our-history.

62 **"dead, bleached, pale, and normal":** Paul L. Jokiel and Stephen L. Coles, "Effects of Heated Effluent on Hermatypic Corals at Kahe Point, Oahu," *Pacific Science* 28, no. 1 (1974): 1–17, https:// scholarspace.manoa.hawaii.edu/bitstream/10125/1144/v28n1-1-18.pdf.

62 **2°C higher:** Stephen L. Coles et al., "Thermal Tolerance in Tropical versus Subtropical Pacific Reef Corals," *Pacific Science* 30, no. 2 (1976): 159–66, https://scholarspace.manoa.hawaii.edu /bitstream/10125/10775/1/v30n2-159-66.pdf.

63 **"Degree Heating Weeks":** "Satellites & Bleaching: 50km Degree Heating Week," Coral Reef Watch, NOAA Satellite and Information Service, https://coralreefwatch.noaa.gov/satellite /education/tutorial/crw24_dhw_product.php.

63 **In the Florida Keys:** Bill Causey, "6. Coral Reefs of the U.S. Caribbean: The History of Massive Coral Bleaching and Other Perturbations in the Florida Keys," *Status of Caribbean Coral Reefs after Bleaching and Hurricanes in 2005* (Townsville: Global Coral Reef Monitoring Network, and Reef and Rainforest Research Centre, 2008): 61-67, https://www.coris.noaa.gov/activities/caribbean _rpt/SCRBH2005_06.pdf.

64 **bleaching went global:** Peter W. Glynn, "Widespread Coral Mortality and the 1982–83 El Niño Warming Event," *Environmental Conservation* 11, no. 2 (1984): 133–46, https://miami.pure.elsevier .com/en/publications/widespread-coral-mortality-and-the-1982-1983-el-ni%C3%B1o -warming-even.

64 **In 1987 it was the Caribbean:** William Steif, "Experts Are Puzzled by Widespread Coral 'Bleaching' in Caribbean," *The New York Times*, December 15, 1987, https://www.nytimes.com /1987/12/15/science/experts-are-puzzled-by-widespread-coral-bleaching-in-caribbean.html.

64 **Maldives and Costa Rica:** Terry P. Hughes et al., Supplementary Materials for "Spatial and Temporal Patterns of Mass Bleaching of Corals in the Anthropocene," *Science* 359, no. 6371 (2018), https://science.sciencemag.org/content/sci/suppl/2018/01/03/359.6371.80.DC1/aan8048 _Hughes_SM.pdf.

64 **in 1991, Thailand and Tahiti:** "Science: Is Coral Bleaching Caused by Global Warming?" *New Scientist*, August 16, 1991, https://www.newscientist.com/article/mg13117823-600-science-is -coral-bleaching-caused-by-global-warming/.

64 **in 1993, Bali . . . in 1994, the Central Pacific:** In Table S1, Hughes et al., Supplementary Materials for "Spatial and Temporal Patterns of Mass Bleaching of Corals in the Anthropocene."

64 **thirty-two countries . . . seven hundred years old:** "ISRS Statement on Global Coral Bleaching in 1997–1998," International Society for Reef Studies, http://coralreefs.org/wp-content /uploads/2019/01/ISRS-Statement-2-Coral-Bleaching-in-1997-98.pdf.

64 **An assessment:** Global Coral Reef Monitoring Network, Clive Wilkinson, ed., "Status of Coral Reefs of the World: 2000" (Townsville: Australian Institute of Marine Science, 2000): 18, https://www.icriforum.org/wp-content/uploads/2019/12/gcrmn2000.pdf.

65 **A new kind of mark:** Jessica E. Carilli et al., "Century-Scale Records of Coral Growth Rates Indicate That Local Stressors Reduce Coral Thermal Tolerance Threshold," *Global Change Biology* 16, no. 4 (2010): 1247–57, https://doi.org/10.1111/j.1365-2486.2009.02043.x.

65 **encircled the globe yet again:** Table S1, Hughes et al., Supplementary Materials for "Spatial and Temporal Patterns of Mass Bleaching of Corals in the Anthropocene."

65 **even without El Niño:** Terry Hughes et al., "Spatial and Temporal Patterns of Mass Bleaching of Corals in the Anthropocene," *Science* 359, no. 6371 (2018): 80–83, https://science.sciencemag.org /content/sci/359/6371/80.full.pdf.

65 **In 2015, another blast:** CBS News and Associated Press, "Coral Bleaching Crisis Spreads Worldwide—and It's Getting Worse," CBSNews.com, October 8, 2015, https://www.cbsnews .com/news/coral-bleaching-crisis-spreads-worldwide/.

65 **By 2016, the warm water:** Table S1, Hughes et al., Supplementary Materials for "Spatial and Temporal Patterns of Mass Bleaching of Corals in the Anthropocene."

65 **the heat returned:** "2017 Marine Heatwave on the Great Barrier Reef," Bureau of Meteorology, Australian Government, http://www.bom.gov.au/environment/doc/marine-heatwave.pdf. Richard Schiffman, "A Close-Up Look at the Catastrophic Bleaching of the Great Barrier Reef," *Yale Environment 360*, April 10, 2017, https://e360.yale.edu/features/inside-look-at-catastrophic -bleaching-of-the-great-barrier-reef-2017-hughes.

66 **corals underwent bleaching:** Hayley Warren, "Half the World's Coral Reefs Already Have Been Killed by Climate Change," Bloomberg, October 11, 2019, https://www.bloomberg.com /graphics/2019-coral-reefs-at-risk/.

66 **"I can't even tell":** Michael Slezak, "The Great Barrier Reef: A Catastrophe Laid Bare," *The Guardian*, June 6, 2016, https://www.theguardian.com/environment/2016/jun/07/the-great -barrier-reef-a-catastrophe-laid-bare.

66 **three quarters of the reefs:** Schiffman, "A Close-Up Look at the Catastrophic Bleaching of the Great Barrier Reef."

66 **more than 90 percent:** Slezak, "The Great Barrier Reef: A Catastrophe Laid Bare." Terry P. Hughes et al., "Global Warming and Recurrent Mass Bleaching of Corals," *Nature* 543 (2017): 373–77, https://doi.org/10.1038/nature21707.

66 **repeated an experiment:** Steve L. Coles et al., "Evidence of Acclimatization or Adaptation in Hawaiian Corals to Higher Ocean Temperatures," *PeerJ* 6, no. 5347 (2018), doi:http://dx.doi.org .ezproxy.lib.utexas.edu/10.7717/peerj.5347.

66 **this kind of adaptation:** S. Sully et al., "A Global Analysis of Coral Bleaching over the Past Two Decades," *Nature Communications* 10, no. 1264 (2019), https://www.nature.com/articles/s41467 -019-09238-2/.

67 **In 1881, German biologist:** Thomas Krueger, "Concerning the Cohabitation of Animals and Algae—An English Translation of K. Brandt's 1881 Presentation 'Ueber das Zusammenleben von Thieren und Algen,'" *Symbiosis* 71 (2017): 167–74, https://doi.org/10.1007/s13199-016-0439-2. Brandt coined the taxonomic names Zoochlorella and Zooxanthella to describe the green and yellow cells, respectively. The "chlor" and "xanth" are from the Greek words for green and yellow. The "-ella" adds a diminutive, expressing the tiny size of the algae, about 5 to 13 microns, not much larger than bacteria. A few years later, Brandt realized that the green and yellow cells were just color variants of the same algae and retracted Zoochlorella, leaving us with the general word *zooxanthellae*.

68 **accessory pigments:** Melissa S. Roth, "The Engine of the Reef: Photobiology of the Coral-Algal Symbiosis," *Frontiers in Microbiology* 5 (2014): 422, https://www.frontiersin.org/articles/10.3389 /fmicb.2014.00422/full.

68 **the word *zooxanthellae*:** Todd C. LeJeunesse, "Zooxanthellae Quick Guide," *Current Biology* 30, no. 19 (2020): PR1110–R1113, https://www.cell.com/current-biology/pdf/S0960-9822(20) 30428-0.pdf.

69 **major taxonomic revision:** Todd C. LaJeunesse et al., "Systematic Revision of Symbiodiniaceae Highlights the Antiquity and Diversity of Coral Endosymbionts," *Current Biology* 28, no. 16 (2018): 2570–80.e6, https://www.sciencedirect.com/science/article/pii/S0960982218309072.

69 **options for a coral:** Ray Berkelmans and Madeleine J. H. van Oppen, "The Role of Zooxanthellae in the Thermal Tolerance of Corals: A 'Nugget of Hope' for Coral Reefs in an Era of Climate Change," *Proceedings of the Royal Society B: Biological Sciences* 273, no. 1599 (2006): 2305–12, http://doi.org/10.1098/rspb.2006.3567. D. Tye Pettay et al., "Microbial Invasion of the Caribbean by an Indo-Pacific Coral Zooxanthella," *Proceedings of the National Academy of Sciences of the United States of America* 112, no. 24 (2015): 7513–18, https://www.pnas.org/content/112/24/7513.

70 **In American Samoa:** Megan K. Morikawa and Stephen R. Palumbi, "Using Naturally Occurring Climate Resilient Corals to Construct Bleaching-Resistant Nurseries," *Proceedings of the National Academy of Sciences of the United States of America* 116, no. 21 (2019): 10586–91, https://www.pnas.org/content/116/21/10586.

Chapter 10. Unmasking Immunity

71 **women earned:** "Women, Minorities, and Persons with Disabilities in Science and Engineering," National Center for Science and Engineering Statistics (NCSES), National Science Foundation, March 8, 2019, https://ncses.nsf.gov/pubs/nsf19304/.

71 **National Academy of Sciences:** Laura Hoopes, "National Academy of Sciences Picks Few Women Again," Women in Science (forum), Scitable by Nature Education, May 21, 2011, https://www.nature.com/scitable/forums/women-in-science/national-academy-of-sciences-picks-few-women-19909665/.

74 **"bleaching every which way":** Evelyn Abbott et al., "Disentangling Coral Stress and Bleaching Responses by Comparing Gene Expression in Symbiotic Partners," *SICB 2020 Abstract Book* (Herndon, VA: Society for Integration and Comparative Biology, 2020), 1, https://burkclients.com/sicb/meetings/2020/site/files/2020SICB_AbstractBook.pdf.

75 **ramped-up immune system:** Katelyn M. Mansfield et al., "Transcription Factor NF-κB Is Modulated by Symbiotic Status in a Sea Anemone Model of Cnidarian Bleaching," *Scientific Reports* 7, no. 16025 (2017), https://www.ncbi.nlm.nih.gov/pubmed/29167511. Katelyn M. Mansfield et al., "Varied Effects of Algal Symbionts on Transcription Factor NF-κB in a Sea Anemone and a Coral: Possible Roles in Symbiosis and Thermotolerance," *bioRxiv*, May 17, 2019, https://www.biorxiv.org/content/10.1101/640177v1.

76 **PANDAS:** "PANDAS—Questions and Answers," National Institute of Mental Health, U.S. Department of Health and Human Services (2019), https://www.nimh.nih.gov/health/publications/pandas/index.shtml.

76 **"cross-reaction":** Madeleine W. Cunningham, "Molecular Mimicry, Autoimmunity, and Infection: The Cross-Reactive Antigens of Group A Streptococci and Their Sequelae," *Microbiology Spectrum* 7, no. 4 (2019), https://www.asmscience.org/content/journal/microbiolspec/10.1128/microbiolspec.GPP3-0045-2018.

77 **blood-brain barrier:** "How Immune Response to Strep Infection Triggers BGE Breakdown of Blood Brain Barrier PANDAS/PANS," PANDAS Network, May 12, 2020, https://www.youtube.com/watch?v=4tb6QhhEUEE. Thamotharampillai Dileepan et al., "Group A Streptococcus Intranasal Infection Promotes CNS Infiltration by Streptococcal-Specific Th17 Cells," *The Journal of Clinical Investigation* 126, no. 1 (2015): 303–17, https://www.jci.org/articles/view/80792. Maryann Platt et al., "Th17 Lymphocytes Drive Vascular and Neuronal Deficits in a Mouse Model of Postinfectious Autoimmune Encephalitis," *Proceedings of the National Academy of Sciences of the United States of America* 117, no. 12 (2020): 6708–16, https://www.pnas.org/content/117/12/6708.

Chapter 11. The Holobiont

79 **artillery of stinging cells:** Debora de Oliveira Pires, "Cnidae of Scleractinia," *Proceedings of the Biological Society of Washington* 110, no. 2 (July 9, 1997): 167–85, https://archive.org/stream/proceedingsofb1101997biol/proceedingsofb1101997biol_djvu.txt. Thomas W. Holstein, "A View to Kill," *BMC Biology* 10, no. 18 (2012), http://www.biomedcentral.com/1741-7007/10/18.

79 **barrier of mucus:** David G. Bourne et al., "Insights into the Coral Microbiome: Underpinning the Health and Resilience of Reef Ecosystems," *Annual Review of Microbiology* 70 (2016): 317–40, https://geodynamicsprogram.whoi.edu/wp-content/uploads/sites/14/2018/10/Coral_Microbiome_268944.pdf. Valentine Neunier et al., "Bleaching Forces Coral's Heterotrophy on Diazotrophs and *Synechococcus*," *The ISME Journal: Multidisciplinary Journal of Microbial Ecology* 13 (2019): 2882–86, https://www.nature.com/articles/s41396-019-0456-2.

79 **gooey biochemical moat:** Ryota Nakajima, "The Role of Coral Mucus 101," RyotaNakajima.com, September 8, 2018, https://ryotanakajima.com/coral-reefs/the-role-of-coral-mucus-101.

80 **that makes vitamin B$_{12}$:** Nathan H. Lents, "The Evolutionary Quirk That Made Vitamin B$_{12}$ Part of Our Diet," *Discover*, August 13, 2018, https://www.discovermagazine.com/health/the-evolutionary-quirk-that-made-vitamin-b12-part-of-our-diet.

80 **unique microbiomes:** Forest Rohwer et al., "Diversity and Distribution of Coral-Associated Bacteria," *Marine Ecology Progress Series* 243 (2002): 10, https://www.int-res.com/articles/meps2002/243/m243p001.pdf.

80 **microbiomes shift:** Maren Ziegler et al., "Coral Bacterial Community Structure Responds to Environmental Change in a Host-Specific Manner," *Nature Communications* 10, no. 3092 (2019), https://www.nature.com/articles/s41467-019-10969-5. Christina Korownyk et al., "Population Level Evidence for Seasonality of the Human Microbiome," *Chronobiology International* 35, no. 4 (January 17, 2018): 573–77, https://www.tandfonline.com/doi/abs/10.1080/07420528.2018.1424718.

82 **probiotic vaccine:** Leah Reshef et al., "The Coral Probiotic Hypothesis," *Environmental Microbiology* 8, no. 12 (2007): 2068–73, https://www.tau.ac.il/lifesci/departments/zoology/members/loya/175.pdf.pdf. Raquel S. Peixoto et al., "Coral Probiotics: Premise, Promise, Prospects," *Annual Review of Animal Bioscience* 16, no. 9 (2021): 265–88, https://pubmed.ncbi.nlm.nih.gov/33321044.

82 **cooperative consortia:** Margulis originally used the term holobiont to describe the badass merger between coral and algae, then recognized its application to so many other systems. Kevin Slavin, "Design and Science of the Holobiont," *Journal of Design and Science* (June 6, 2017), https://jods.mitpress.mit.edu/pub/design-science-holobiont/release/1. Lynne Margulis and René Fester, *Symbiosis as a Source of Evolutionary Innovation: Speciation and Morphogenesis* (Cambridge, MA: MIT Press, 1991).

82 **holobiont:** Eugene Rosenberg and Ilana Zilber-Rosenberg, "The Hologenome Concept of Evolution after 10 Years," *Microbiome* 6, no. 78 (2018), https://doi.org/10.1186/s40168-018-0457-9.

Chapter 12. Throwing Shade

Thank you to my fabulous summer intern, Annie Ferdman, for research for this chapter.

85 **The two most cited:** C. A. Downs et al., "Toxicopathological Effects of the Sunscreen UV Filter, Oxybenzone (Benzophenone-3), on Coral Planulae and Cultured Primary Cells and Its Environmental Contamination in Hawaii and the U.S. Virgin Islands," *Archives of Environmental Contamination and Toxicology* 70, no. 2 (2016): 265–88, https://pubmed.ncbi.nlm.nih.gov/26487337/. Craig Downs et al., "Toxicological Effects of the Sunscreen UV Filter, Benzophenone-2, on Planulae and In Vitro Cells of the Coral, *Stylophora pistillata*," *Ecotoxicology* 23, no. 2 (2014): 175–91, https://pubmed.ncbi.nlm.nih.gov/24352829/.

86 **Replicate work:** Tangtian He et al., "Comparative Toxicities of Four Benzophenone Ultraviolet Filters to Two Life Stages of Two Coral Species," *Science of the Total Environment* 651, no. 2 (2019): 2391–99, https://doi.org/10.1016/j.scitotenv.2018.10.148. Tim Wijgerde et al., "Adding Insult to Injury: Effects of Chronic Oxybenzone Exposure and Elevated Temperature on Two Reef-Building Corals," *Science of the Total Environment* 733 (2020): 139030, https://doi.org/10.1016/j.scitotenv.2020.139030.

86 **review of two dozen studies:** Elizabeth Wood, "Impacts of Sunscreens on Coral Reefs," *ICRI Briefing*, International Coral Reef Initiative (February 2018), https://www.icriforum.org/wp-content/uploads/2019/12/ICRI_Sunscreen_0.pdf.

87 **what chemicals in sunscreens do:** Margaret Schlumpf et al., "In Vitro and In Vivo Estrogenicity of UV Screens," *Environmental Health Perspectives* 109, no. 3 (2001): 239–44, https://pubmed.ncbi.nlm.nih.gov/11333184/. Steven Wang et al., "Safety of Oxybenzone: Putting Numbers into Perspective," *Archives of Dermatology* 147, no. 7 (2011): 865–66, https://jamanetwork.com/journals/jamadermatology/fullarticle/1105240. Marya Ghazipura et al., "Exposure to Benzophenone-3 and Reproductive Toxicity: A Systematic Review of Human and Animal Studies,"

Reproductive Toxicology 73 (2017): 175–83, https://www.sciencedirect.com/science/article/pii /S0890623817302277.

87 **oxybenzone in nearly everyone:** Antonia M. Calafat et al., "Concentrations of the Sunscreen Agent Benzophenone-3 in Residents of the United States: National Health and Nutrition Examination Survey 2003–2004," *Environmental Health Perspectives* 116, no. 7 (2008): 893–97, https://ehp.niehs.nih.gov/doi/10.1289/ehp.11269.

87 **more absorbable:** Murali K. Matta et al., "Effect of Sunscreen Application on Plasma Concentration of Sunscreen Active Ingredients: A Randomized Clinical Trial," *JAMA* 323, no. 3 (2020): 256–67, https://jamanetwork.com/journals/jama/article-abstract/2759002.

87 **triggered a requirement:** "Sunscreen Drug Products for Over-the-Counter Human Use," a proposed rule by the Food and Drug Administration, Department of Health and Human Services, February 26, 2019, https://www.federalregister.gov/documents/2019/02/26/2019 -03019/sunscreen-drug-products-for-over-the-counter-human-use. Amanda Turney, an officer at the FDA, responding to my email, wrote that fifteen hundred comments, including "significant amounts of substantive data," were submitted on the rule change during the open period that concluded June 27, 2019.

87 **In Europe:** "Does Europe Have Better Sunscreens?," EWG's Guide to Sunscreens, Environmental Working Group, 2021, https://www.ewg.org/sunscreen/report/does-europe-have-better -sunscreens/.

88 **mycosporine amino acids:** Melissa Pandika, "Looking to Nature for New Sunscreens," *American Chemical Society Central Science* 4, no. 7 (2018): 788–90, https://pubs.acs.org/doi/10.1021 /acscentsci.8b00433. Karl P. Lawrence, "Mycosporine-Like Amino Acids for Skin Photoprotection," *Current Medicinal Chemistry* 25, no. 40 (December 2018): 5512–27, https://pubmed.ncbi .nlm.nih.gov/28554325/.

88 **CARES Act:** Genevieve M. Razick, "It Is Not All about the Coronavirus: The CARES Act Brings Long-Awaited Over-the-Counter (OTC) Monograph Reform," Arnall Golden Gregory LLP, April 10, 2020, https://www.agg.com/news-insights/publications/it-is-not-all-about-the -coronavirus-the-cares-act-brings-long-awaited-over-the-counter-otc-monograph-reform/.

88 **ban the sale of sunscreens:** Shannon McMahon, "6 Destinations with Sunscreen Bans, and What You Need to Know," Smarter Travel, August 20, 2019, https://www.smartertravel.com /sunscreen-ban-destinations/. Lena Reece, "Leleuvia Island Resort Bans Chemical Sunscreens on the Island," *Fiji Village*, July 31, 2018, https://fijivillage.com/news/Leleuvia-Island-Resort -bans-chemical-sunscreens-on-the-island-r2s95k. Matt McGrath, "Coral: Palau to Ban Sunscreen Products to Protect Reefs," BBC News, November 1, 2018, https://www.bbc.com/news /science-environment-46046064. The Hawai`i ban went into effect on January 1, 2021. Stephanie Pappas, "Another Tropical Paradise Enacts a Sunscreen Ban," *Live Science*, May 17, 2018, https://www.livescience.com/62598-bonaire-island-bans-sunscreen.html. The Florida ban was scheduled for July 1, 2021, although it was challenged in court. Jim Turner, "Senate Moves Quickly to Stop Sunscreen Bans," *South Florida Sun Sentinel*, January 15, 2020, https://www.sun -sentinel.com/news/florida/fl-ne-florida-sunscreen-rules-20200115-baoxpnpl5ff75hm5k 7kl5jwmda-story.html.

89 **"reef safe" products:** "Unsubstantiated Coral 'Reef Safe' Claims Enjoined in California," Office of the District Attorney Jill Ravitch, County of Sonoma, California, October 7, 2020, https:// patch.com/california/sonomavalley/unsubstantiated-coral-reef-safe-claims-enjoined -california.

Chapter 13. Fragmentation

93 **Around two hundred million years ago:** Adam Skarke, "Geologic Overview of the Gulf of Mexico," NOAA 2018 Gulf of Mexico Expedition, NOAA Ocean Exploration, National Oceanic and Atmospheric Administration, U.S. Department of Commerce, 2018, https://ocean explorer.noaa.gov/okeanos/explorations/ex1803/background/geology/welcome.html.

93 **But around 1950, the coral thickets:** Katie L. Cramer et al., "Widespread Loss of Caribbean Acroporid Corals Was Underway before Coral Bleaching and Disease Outbreaks," *Science Advances* 6, no. 17 (2020): eaax9395, https://advances.sciencemag.org/content/6/17/eaax9395.full.

94 **white band disease:** David I. Kline and Steven V. Vollmer, "White Band Disease (Type I) of Endangered Caribbean Acroporid Corals Is Caused by Pathogenic Bacteria," *Scientific Reports* 1, no. 7 (June 14, 2011), https://doi.org/10.1038/srep00007.

94 **sea urchins . . . parrotfishes:** H. A. Lessios, "The Great *Diadema antillarum* Die-Off: 30 Years Later," *Annual Review of Marine Science* 8 (2016): 267–83, https://doi.org/10.1146/annurev-marine -122414-033857. Katie L. Cramer et al., "Prehistorical and Historical Declines in Caribbean Coral Reef Accretion Rates Driven by Loss of Parrotfish," *Nature Communications* 8, no. 14160 (2017), https://www.nature.com/articles/ncomms14160.

94 **2 percent of the original:** Hannah Chinn, "'Dire Outlook': Scientists Say Florida Reefs Have Lost Nearly 98% of Coral," *The Guardian*, November 18, 2020, https://www.theguardian.com /environment/2020/nov/18/coral-reefs-florida-dire-outlook.

94 **Dave Vaughan:** Plant a Million Corals Foundation, http://plantamillioncorals.org.

94 **largest ornamental fish companies:** Oceans, Reefs & Aquariums (ORA), https://www.ora farm.com/.

95 **YouTube interview:** "Dr. David Vaughan—Restoring 100-Year-Old Coral in Just Two Years," Reef Restoration and Adaptation Program, October 3, 2018, https://www.youtube.com/watch?v =e7n19msJhmQ.

95 **Ken Nedimyer:** Jessica Weiss, "Meet the Florida Man Replanting Thousands of Corals to Fight a Historic Die-Off," *Miami New Times*, December 11, 2015, https://www.miaminewtimes.com /news/meet-the-florida-man-replanting-thousands-of-corals-to-fight-a-historic-die-off -8107654. Lan Trinh, "Florida Group Rebuilds Vital Coral Reefs," CNN.com, updated March 1, 2012, https://www.cnn.com/2012/03/01/us/cnnheroes-nedimyer-coral-reefs/index.html. Coral Restoration Foundation, https://www.coralrestoration.org/about.

97 **"Eureka mistake":** "The Eureka Mistake," Plant a Million Corals Foundation, July 25, 2019, http://plantamillioncorals.org/the-eureka-mistake/.

97 **"sweeper tentacles":** R. B. Williams, "Acrorhagi, Catch Tentacles and Sweeper Tentacles: A Synopsis of 'Aggression' of Actiniarian and Scleractinian Cnidaria," *Hydrobiologia* 216 (1991): 539–45, https://link.springer.com/article/10.1007%2FBF00026511.

98 **fusing or reskinning:** Christopher A. Page et al., "Microfragmenting for the Successful Restoration of Slow Growing Massive Corals," *Ecological Engineering* 123 (2018): 86–94, https://www. sciencedirect.com/science/article/pii/S0925857418303094.

98 **Mote Marine Lab:** *Annual Report 2018*, Mote Marine Lab, p. 23, https://mote.org/media /uploads/files/2018AnnualReport_ffw.pdf. *Annual Report 2020*, Mote Marine Lab, 13, https:// mote.org/media/uploads/files/2020AnnualReport_ffw_spreads_updated.pdf.

98 **Coral Restoration Foundation:** Coral Restoration Foundation, https://www.coralrestoration .org/restoration. "NOAA Backs Coral Restoration Foundation[TM] with More Than $2.5 Million to Restore Florida's Coral Reefs," Accesswire, September 16, 2019, https://www.accesswire .com/559798/NOAA-Backs-Coral-Restoration-FoundationTM-with-more-than-25-Million -to-Restore-Floridas-Coral-Reefs.

Chapter 14. Outbreak

99 **Panama Canal:** "The Expanded Canal," *Canal de Panamá*, May 15, 2019, https://micanaldepanama .com/expansion/. "Panama Canal: Expansion into the 21st Century," WSP.com, 2021, https:// www.wsp.com/en-GL/projects/panama-canal-expansion.

99 **In Miami:** Chabeli Herrera, "Despite Recent Dredge, PortMiami Still Can't Fit Some Large Ships. New Project in the Works," *The Miami Herald*, July 8, 2018, https://www.miamiherald .com/news/business/article214376334.html.

99 **Waterkeeper Alliance:** Waterkeeper Alliance, https://waterkeeper.org/.

100 **corals near the Port of Miami:** Ross Cunning et al., "Extensive Coral Mortality and Critical Habitat Loss Following Dredging and Their Association with Remotely-Sensed Sediment Plumes," *Marine Pollution Bulletin* 145 (2019): 185–99, https://d3n8a8pro7vhmx.cloudfront.net /miamiwaterkeeper/pages/3099/attachments/original/1589896592/Cunning_et_al _Published_Manuscript.pdf?1589896592. "Lawsuit over Dredging Achieves Restoration of 10,000 Threatened Corals," *Portmiami Settlement E-News*, Miami Waterkeeper, August 6, 2018, https://www.miamiwaterkeeper.org/victory_alert_portmiami_legal_settlement. Manuel Madrid, "Five Ways Miami Screwed the Environment in 2019," *Miami New Times,* December 26, 2019, https://www.miaminewtimes.com/news/five-ways-miami-screwed-the-environment-in-2019 -11353111. Jerry Iannelli, "PortMiami Dredging Killed a Half-Million Corals, New Study Shows," *Miami New Times*, May 29, 2019, https://www.miaminewtimes.com/news/portmiami-dredging -killed-a-half-million-corals-miami-waterkeeper-study-says-11184042.

101 **webinar about the spread:** "Stony Coral Tissue Loss Disease Florida," Reef Resilience, May 8, 2019, https://youtu.be/Qprs7m1tGPA.

102 **a map of the Caribbean:** "Florida's Coral Disease Outbreak," Florida Keys National Marine Sanctuary, National Ocean Service, National Oceanic and Atmospheric Administration, 2021, https://floridakeys.noaa.gov/coral-disease/. L. Roth et al., "Caribbean SCTLD Dashboard," ArcGIS Online, accessed July 6, 2021, https://www.agrra.org/coral-disease-outbreak/.

103 **"reefugees" of the Caribbean:** Jennifer Kay, "Mystery Disease Killing Florida Corals Spurs 'Reef-Ugee' Mission," Bloomberg, May 20, 2019, https://news.bloombergenvironment.com /environment-and-energy/mystery-disease-killing-florida-corals-spurs-reef-ugee-mission.

103 **joined together to fight:** "Stony Coral Tissue Loss Disease Response," Florida Department of Environmental Protection, April 1, 2021, https://floridadep.gov/rcp/coral/content/stony -coral-tissue-loss-disease-response.

103 **valuing the Florida reef:** "Coral Reef Barriers Provide Flood Protection for More Than 18,000 People and $1.8 Billion Worth of Coastal Infrastructure and Economic Activity Annually," U.S. Geological Survey, April 30, 2019, https://www.usgs.gov/news/coral-reef-barriers-provide-flood -protection-more-18000-people-and-18-billion-worth-coastal.

104 **economic activities:** "Florida," Coral Reef Information System (CoRIS), NOAA, March 22, 2021, www.coris.noaa.gov/portals/florida.html.

104 **tank I'd asked to see:** Taylor Torregano, "Mote Cares for Corals Rescued in Front of the Path of Deadly Disease," WWSB News, June 14, 2019, https://www.mysuncoast.com/2019/06/13 /mote-cares-corals-rescued-front-path-deadly-disease/.

Chapter 15. Lesion by Legion

106 **EarthX:** Peter Simek, "EarthX Turned Trammell S. Crow into an Unlikely Environmentalist," *D Magazine*, April 2018, https://www.dmagazine.com/publications/d-magazine/2018/april /trammell-s-crow-earthx-earth-day/.

107 **expanded the protected area:** "Expansion Proposal for Flower Garden Banks Sanctuary Released," National Oceanic and Atmospheric Administration, May 1, 2020, https://www.noaa.gov /media-release/expansion-proposal-for-flower-garden-banks-sanctuary-released. "DEIS for Sanctuary Expansion," Flower Garden Banks National Marine Sanctuary, National Ocean Service, NOAA, January 19, 2021, https://flowergarden.noaa.gov/management/expansiondeis .html.

107 **"There used to be a bleaching":** "Coral Bleaching," Flower Garden Banks National Marine Sanctuary, National Ocean Service, NOAA, January 19, 2021, https://flowergarden.noaa.gov /education/bleaching.html.

107 **"Force Blue":** Force Blue, https://forceblueteam.org/.

109 **National Football League:** "100 Yards of Hope Team Outplants 1100 Corals," Force Blue, November 9, 2020, https://forceblueteam.org/news/100-yards-of-hope-team-outplants-1100 -corals/.

Chapter 16. The Tragedy of Scale

115 **"The Tragedy of the Commons":** Originally given as an address to the Pacific Division of the American Association for the Advancement of Science, then republished as Garrett Hardin, "The Tragedy of the Commons," *Science* 162, no. 3859 (1968): 1243–48, https://science.sciencemag.org /content/162/3859/1243.full.

116 **California's sardines:** Patricia Wolf, "Recovery of the Pacific Sardine and the California Sardine Fishery," *CalCOFI Report* 33 (1992): 76–86, http://www.calcofi.org/publications/calco fireports/v33/Vol_33_Wolf.pdf.

116 **In 1992, the cod fishery:** Kenneth T. Frank et al., "Trophic Cascades in a Formerly Cod-Dominated Ecosystem," *Science* 308, no. 5728 (2005): 1621–23, https://www.jstor.org/stable /3841620.

117 **That same year, red snapper:** "History of Management of Gulf of Mexico Red Snapper," accessed March 23, 2021, NOAA Fisheries, https://www.fisheries.noaa.gov/history-management -gulf-mexico-red-snapper.

117 **orange roughy:** P. L. Cordue, "Orange Roughy: The Story of New Zealand Orange Roughy from the 'Poster Child' of Unsustainable Fishing to Marine Stewardship Council Assessment," Sustainable Fisheries, https://sustainablefisheries-uw.org/fishery-feature/orange-roughy/.

117 **Pacific salmon:** Eben Harrell, "What Is Killing the Pacific Salmon?," *Time*, January 14, 2011, https://science.time.com/2011/01/14/what-is-killing-the-pacific-salmon/.

117 **seas might have limits:** H. Scott Gordon, "The Economic Theory of a Common-Property Resource: The Fishery," *The Journal of Political Economy* 62, no. 2 (1954): 124–42, http://www.jstor .org/stable/1825571.

117 **Lizzie Mcleod:** "Lizzie Mcleod," The Nature Conservancy, https://www.nature.org/en-us /about-us/who-we-are/our-people/lizzie-mcleod/.

117 **great center of biodiversity:** Muhammad Abrar et al., "Status Terumbu Karang Indonesia 2017," Pusat Penelitian Oseanografi—Lembaga Ilmu Pengetahuan Indonesia, 2017: 11, using Google Translate. Danwei Huang et al., "The Origin and Evolution of Coral Species Richness in a Marine Biodiversity Hotspot," *Evolution* 72, no. 2 (February 2018): 288–302, http://eeg .github.io/lab/_static/reprints/coral_triangle.pdf.

118 **rules of *sasi*:** Elizabeth Mcleod et al., "*Sasi* and Marine Conservation in Raja Ampat, Indonesia," *Coastal Management* 37, no. 6 (2009): 656–76, https://doi.org/10.1080/08920750903244143.

118 **"traditional marine tenure":** Robert E. Johannes, "Traditional Marine Conservation Methods in Oceania and Their Demise," *Annual Review of Ecology and Systematics* 9 (1978): 349–64, https:// www.jstor.org/stable/2096753.

118 **marine protected areas:** "The Marine Protection Atlas," Marine Conservation Institute, 2020, http://mpatlas.org, cited in Enric Sala et al., "Protecting the Global Ocean for Biodiversity, Food and Climate," *Nature* 592 (2021): 397–402, https://www.nature.com/articles/s41586 -021-03371-z.

119 **"paper MPAs":** Peter Howson, "A Huge Marine Reserve in the Pacific Will Protect Rich Tour- ists Rather Than Fish," *The Conversation*, November 1, 2017, https://theconversation.com/a -huge-marine-reserve-in-the-pacific-will-protect-rich-tourists-rather-than-fish-85770. Kirsten Grorud-Colvert and Jane Lubchenco, "Momentum Grows for Ocean Preserves. How Well Do They Work?," *The Conversation*, January 3, 2017, https://theconversation.com/momentum -grows-for-ocean-preserves-how-well-do-they-work-65625?xid=PS_smithsonian.

119 **30x30:** "How to Save Our Ocean #30by30—Narrated by Sting," Department for Environment, Food and Rural Affairs (UK Government), January 16, 2020, https://www.youtube.com /watch?v=4wOH_HhY_Yk&feature=youtu.be. "Zero Draft of the Post-2020 Global Biodiver- sity Framework," Convention on Biological Diversity, January 9, 2020, https://www.cbd.int /doc/c/efb0/1f84/a892b98d2982a829962b6371/wg2020-02-03-en.pdf.

120 **Elinor Ostrom:** "Elinor Ostrom, Nobel Laureate, 1933–2012," Indiana University, https:// www.elinorostrom.com/.

121 **"drama of the commons":** Eduardo Araral, "Ostrom, Hardin and the Commons: A Critical Appreciation and a Revisionist View," *Environmental Science & Policy* 36 (2014): 11–23, https:// correctphilippines.org/wp-content/uploads/2020/06/s03-Araral_2014_commons.pdf. Elinor Ostrom et al., eds., *The Drama of the Commons* (Washington, DC: National Academy Press, 2002), https://doi.org/10.17226/10287.

121 **A 2018 study of twenty-seven MPAs:** Sylvaine Giakoumi et al., "Revisiting 'Success' and 'Fail- ure' of Marine Protected Areas: A Conservation Scientist Perspective," *Frontiers in Marine Science* 5 (2018): 223, https://www.frontiersin.org/articles/10.3389/fmars.2018.00223/full.

Chapter 17. The Scale of Tragedy

122 **Mars, Incorporated:** "Mars," *Forbes*, 2021, https://www.forbes.com/companies/mars/#55efa 2f83bb7. Hillary Hoffower, "Meet the Mars Family, Heirs to the Snickers and M&M's Candy Empire, Who Spent Years Avoiding the Limelight and Are America's Third-Wealthiest Family 'Dynasty,'" *Business Insider*, March 27, 2019, https://www.businessinsider.com/mars-inc-family -fortune-net-worth-lifestyle-snickers-twix-2019-3. Joel Glenn Brenner, "Life on Mars: The Mars Family Saga Has All the Classic Elements," *The Independent*, October 22, 2011, https://www .independent.co.uk/arts-entertainment/life-on-mars-the-mars-family-saga-has-all-the-classic -elements-1535722.html.

123 **Skittles, Snickers, and Starburst:** Audrey Castoreno, "A Look Inside the Mars Wrigley Con- fectionery in Waco," KWTX-TV, November 7, 2018, https://www.kwtx.com/content/news /A-look-inside-the-Mars-Wrigley-Confectionery-500002791.html.

124 **undersea web:** "Mars Coral Reef Restoration Efforts Show Remarkable Progress," Mars, In- corporated, June 5, 2020, https://www.youtube.com/watch?v=MKPA3R5izT4.

126 **questions of scale:** "Announcing the Release of the World Atlas of Coral Reefs," UNEP Coral
Reef Unit, United Nations Environment Programme World Conservation Monitoring Centre
(UNEP-WCMC), September 11, 2001, http://coral.unep.ch/atlaspr.htm. Estimate is 28,430,000
hectares.

126 **$1 to $2 per coral:** Susan L. Williams et al., "Large-Scale Coral Reef Rehabilitation after Blast
Fishing in Indonesia," *Restoration Ecology* 27, no. 2 (March 2019): 447–56, https://onlinelibrary
.wiley.com/doi/full/10.1111/rec.12866.

126 **The EPA's entire annual budget:** "EPA's Budget and Spending," United States Environmental Protection Agency, https://www.epa.gov/planandbudget/budget.

126 **subsidies paid:** Matthias Kalkuhl et al., "Chapter 6. Bridging the Gap: Fiscal Reforms for
the Low-Carbon Transition," *Emissions Gap Report 2018*, United Nations Environment Programme, November 2018, p. 44. "Subsidies for oil, natural gas and coal amounted to US$373
billion [€319 billion] in 2015." http://wedocs.unep.org/bitstream/handle/20.500.11822/26895
/EGR2018_FullReport_EN.pdf?sequence=1&isAllowed=y. David Coady et al., "How Large Are
Global Fossil Fuel Subsidies?," *Science Direct* 91 (2017): 11–27, https://www.sciencedirect.com
/science/article/pii/S0305750X16304867. "Estimated subsidies are $4.9 trillion worldwide in
2013 and $5.3 trillion in 2015."

126 **$2.4 trillion per year:** Valérie Masson-Delmotte et al., eds., "IPCC, 2018: Summary for Policy-
makers," in *Global Warming of 1.5°C: An IPCC Special Report on the Impacts of Global Warming of 1.5°C
above Pre-Industrial Levels and Related Global Greenhouse Gas Emission Pathways, in the Context of
Strengthening the Global Response to the Threat of Climate Change, Sustainable Development, and Efforts to
Eradicate Poverty* (Geneva: World Meteorological Organization, 2018), 24, https://report.ipcc.ch
/sr15/pdf/sr15_spm_final.pdf.

126 **"ecosystem services":** Robert Costanza et al., "Changes in the Global Value of Ecosystem
Services," *Global Environmental Change* 26 (2014): 152–58, https://www.sciencedirect.com/science
/article/abs/pii/S0959378014000685.

Chapter 18. A Place to Restore

128 **Sulawesi:** An old name for Sulawesi is Celebes. It might have come from Portuguese navigators
who called the dangerous capes in the northeast "the celebrated ones." Because the name Celebes
is associated with the brutal period of Dutch colonialism, it is rarely used anymore.

129 **antimacassar:** Thrifty Traveller, "How Ironic—There's No Macassar Oil in Makassar Any
Longer," *Free Malaysia Today*, November 27, 2018, https://www.freemalaysiatoday.com/category
/leisure/2018/11/27/how-ironic-theres-no-macassar-oil-in-makassar-any-longer/.

131 **Indonesian Through Flow:** Jane Sprintall et al., "Direct Estimates of the Indonesian Through-
flow Entering the Indian Ocean: 2004–2006," *Journal of Geophysical Research: Oceans* 114, no. C7
(2009): 241, https://agupubs.onlinelibrary.wiley.com/doi/pdf/10.1029/2008JC005257.

132 **Indo-Pacific warm pool:** Patrick De Deckker, "The Indo-Pacific Warm Pool: Critical to World
Oceanography and World Climate," *Geoscience Letters* 3, no. 20 (2016), https://geoscienceletters
.springeropen.com/articles/10.1186/s40562-016-0054-3. John Weier, "Reverberations of the
Pacific Warm Pool," NASA Earth Observatory, July 24, 2001, https://earthobservatory.nasa
.gov/features/WarmPool.

132 **surveyed 226 reefs:** Timothy R. McClanahan et al., "Large Geographic Variability in the Resistance of Corals to Thermal Stress," *Global Ecology and Biogeography* 29, no. 12 (2020): 2229–47,
https://onlinelibrary.wiley.com/doi/epdf/10.1111/geb.13191.

Chapter 19. A Reef of Hope

134 **was Wallace's line:** Alfred Russel Wallace, "On the Physical Geography of the Malay Archipelago," reprinted in *Proceedings of the Royal Geographical Society* 7 (1845): 205–12, https://archive
.org/details/proceedingsmonth07royauoft/page/n217/mode/2up?view=theater. Ernst Mayr,
"Wallace's Line in the Light of Recent Zoogeographic Studies," *The Quarterly Review of Biology* 19,
no. 1 (1944), https://www.journals.uchicago.edu/doi/10.1086/394684.

134 **Tuna Blueprint:** "A Vision for Recapturing the Wealth of Tuna," *WWF Factsheet*, 2016, http://
wwf.panda.org/knowledge_hub/where_we_work/coraltriangle/solutions/fisheries/sustainable
_tuna_fisheries_coraltriangle/. "Targeting Tuna," 2018, United Nations Environment Programme,
https://www.unep.org/news-and-stories/story/targeting-tuna.

Chapter 20. A Blast in Makassar

139 **blast fishermen in Sulawesi:** Lida Pet-Soede and Mark V. Erdmann, "Blast Fishing in South-west Sulawesi, Indonesia," *Naga: The ICLARM Quarterly* 21, no. 2 (1998): 4-9, https://www.researchgate.net/publication/227642073_Blast_fishing_in_SW_Sulawesi_Indonesia.

140 **reporting on blast fishing:** Jani Hill, "Watch Fishermen Bomb Their Catch Out of the Water," *National Geographic,* June 3, 2016, https://www.nationalgeographic.com/news/2016/06/blast-fishing-dynamite-fishing-tanzania/. Aurora Almendral, "In the Philippines, Dynamite Fishing Decimates Entire Ocean Food Chains," *The New York Times,* June 15, 2018, https://www.nytimes.com/2018/06/15/world/asia/philippines-dynamite-fishing-coral.html. Malaka Rodrigo, "Crackdown after Sri Lanka Bombings May Help in Fight against Blast Fishing," *Mongabay*, June 26, 2019, https://news.mongabay.com/2019/06/crackdown-after-sri-lanka-bombings-may-help-in-fight-against-blast-fishing/.

141 **Blast fishing was practiced:** Muhammad Chozin, *Illegal but Common* (Saarbrücken, Germany: VDM Verlag Dr. Muller, 2009). Pet-Soede and Erdmann, "Blast Fishing in Southwest Sulawesi, Indonesia."

141 **why does it continue?:** Efforts to decrease the use of bombs using acoustic sensors have been implemented in Malaysia and seem to be working. Elizabeth Wood, "Fighting Fish Bombing with High Tech," Save Our Seas Foundation, July 23, 2015, https://saveourseas.com/update/fighting-fish-bombing-with-high-tech/. The COVID-19 crisis, which quashed the tourist industry in the Coral Triangle, threatened to lead to a resurgence of blast fishing in places where the monitoring had been effective. Siew Lyn Wong, "Tackling Fish-Bombing among the Coral Reefs of Malaysia's Sabah," *China Dialogue Ocean*, March 30, 2021, https://chinadialogueocean.net/16589-fish-bombing-coral-reefs-sabah-malaysia/.

142 **"problem with cyanide":** Christie Wilcox, "Fishing with Cyanide," *Hakai Magazine*, June 30, 2016, https://www.hakaimagazine.com/news/fishing-cyanide/. Rachael Bale, "The Horrific Way Fish Are Caught for Your Aquarium—with Cyanide," *National Geographic*, March 10, 2016, https://www.nationalgeographic.com/news/2016/03/160310-aquarium-saltwater-tropical-fish-cyanide-coral-reefs/.

Chapter 21. Reef Stars

145 **seventy species of damselfish:** Susan Williams et al., "Large-Scale Coral Reef Rehabilitation after Blast Fishing in Indonesia," *Restoration Ecology* 27, no. 2 (2018): 447–56, https://onlinelibrary.wiley.com/doi/full/10.1111/rec.12866.

145 **marine acoustics:** Timothy Gordon, University of Exeter, "Sounds of the Past Give New Hope for Coral Reef Restoration," YouTube, November 29, 2019, https://www.youtube.com/watch?v=dRJeBIu9LZE. Timothy A. C. Gordon et al., "Acoustic Enrichment Can Enhance Fish Community Development on Degraded Coral Reef Habitat," *Nature Communications* 10, no. 5414 (2019), https://www.nature.com/articles/s41467-019-13186-2.

Chapter 22. Galaxy of Potential

155 **the word HOPE:** "The Making of Sheba Hope Reef, Behind the Scenes, Sheba Hope Grows," SHEBA® Brand, YouTube, May 18, 2021, https://www.youtube.com/watch?v=AK_abl7Pm00&t=5s.

Chapter 23. The Coral Cloud

159 **marine aquarium hobby:** Colette Wabnitz et al., *From Ocean to Aquarium: The Global Trade in Marine Ornamental Species* (Cambridge, UK: United Nations Environment Programme World Conservation Monitoring Centre, 2003), https://www.unenvironment.org/resources/report/ocean-aquarium-global-trade-marine-ornamental-species.

160 **rate toxicity:** Christie Wilcox, *Venomous: How Earth's Deadliest Creatures Mastered Biochemistry* (New York: Scientific American/Farrar, Straus and Giroux, 2016), 25–26. It's worth pointing out that the method of delivery of venom is very different for zoanthids, snakes, and spiders, so the risk of harm isn't exactly the same as the LD (lethal dose), which is tested on cells in culture.

161 **manufacture palytoxin:** Vítor Ramos and Vítor Vasconcelos, "Palytoxin and Analogs: Biological and Ecological Effects," *Marine Drugs* 8, no. 7 (2010): 2021–37, http://ncbi.nlm.nih.gov/pmc/articles/PMC2920541/.

162 **international patterns and routes:** Wabnitz et al., *From Ocean to Aquarium.* Allan Craig et al., *Review of Trade in Ornamental Coral, Coral Products and Reef Associated Species to the United States* (Washington, DC: World Wildlife Fund, 2012), https://c402277.ssl.cf1.rackcdn.com/publications/1060/files/original/WWF_Coral_Trade_Report_Final.pdf?1495637324.

162 **illegally harvested wild coral:** E. P. Green and H. Hendry, "Is CITES an Effective Tool for Monitoring Trade in Corals?" *Coral Reefs* 18 (1999): 403–7, https://doi.org/10.1007/s003380050218.

163 **two decades of CITES:** Andrew L. Rhyne, "Long-Term Trends of Coral Imports into the United States Indicate Future Opportunities for Ecosystem and Societal Benefits," *Conservation Letters* 5 (2012): 478–85, https://docs.rwu.edu/fcas_fp/137/. Andre Rhyne et al., Marine Aquarium Diversity and Trade Flow, 2015, https://www.aquariumtradedata.org/.

164 **The moratorium:** Jake Adams, "Fiji Coral & Live Rock BANNED by Ministry of Fisheries!," Reef Builders, December 29, 2017, https://reefbuilders.com/2017/12/29/fiji-coral-live-rock-banned-by-ministry-of-fisheries/. Jake Adams, "Fiji Rock & Coral Ban Officially Withdrawn," Reef Builders, January 31, 2018, https://reefbuilders.com/2018/01/31/fiji-rock-coral-ban-officially-withdrawn/.

164 **new ban on coral exports:** Lawrence Lilley, "Coral Exports Ban: Threat or Opportunity for Sustainability?," *The Jakarta Post,* June 26, 2018, https://www.thejakartapost.com/life/2018/06/26/coral-exports-ban-threat-or-opportunity-for-sustainability.html.

165 **Ocean Gardener:** Nicole Helgason, "The Indo Ban Is Sill On, but Corals Are Growing Out Nicely," Reef Builders, February 9, 2019, https://reefbuilders.com/2019/02/09/indo-ban-corals-farm-growing-nicely. Sabine Penisson, "Visite de la ferme de culture de coraux d'Amblard à Serangan," *CanalBlog,* August 25, 2012, http://lebacasab.canalblog.com/archives/2012/08/25/24962041.html. Ocean Gardener, https://oceangardener.org/.

165 **articles for Reef Builders:** Vincent Chalias, "Alveopora Are NOT a Flowerpot Coral After All!," Reef Builders, February 21, 2019, https://reefbuilders.com/2019/02/21/alevopora-are-not-so-flowerpot-after-all/. Vincent Chalias, "Euphyllia cristata, the Coral Everyone Forgets About," Reef Builders, February 13, 2019, https://reefbuilders.com/2019/02/13/euphyllia-cristata-the-forgotten-coral/.

Chapter 24. The Coral Farm

167 **800 or so species:** Andreas Dietzel et al., "The Population Sizes and Global Extinction Risk of Reef Building Coral Species at Biogeographic Scales," *Nature Ecology & Evolution* 5 (2021): 663–69, https://doi.org/10.1038/s41559-021-01393-4.

169 **mimic algae:** Nicole Helgason, "This Mimic Algae Looks Just Like Branching Acropora Coral!," Reef Builders, March 19, 2018, https://reefbuilders.com/2018/03/19/algae-mimic-acropora/.

170 **Marine Aquarium Conference:** Jake Adams, "Vincent Chalias to Give Important Keynote at MACNA 2019 Banquet," Reef Builders, July 22, 2019, https://reefbuilders.com/2019/07/22/vincent-chalias-to-give-important-keynote-at-macna-2019-banquet/.

171 **Indonesian moratorium on the trade:** Toan Dao, "New Indonesian Fisheries Minister Taking Steps to Wipe Out Predecessor's Controversial Policies," *SeafoodSource,* December 24, 2019, https://www.seafoodsource.com/news/environment-sustainability/new-indonesian-fisheries-minister-taking-steps-to-wipe-out-predecessor-s-controversial-polici. Vincent Chalias, "The First Official Statement on the Coral Ban from the Indonesian Fisheries Dept.," Reef Builders, November 13, 2019, https://reefbuilders.com/2019/11/13/the-first-official-statement-on-the-coral-ban-from-the-indonesian-fisheries-dept/. Vincent Chalias, "Indonesian Cultured Coral Exports Will Resume Next Week!," Reef Builders, January 9, 2020, https://reefbuilders.com/2020/01/09/indonesian-cultured-corals-are-back/.

171 **new bureaucracy:** In a follow-up email in 2021, Vincent added, "Nothing really changed from before . . . Hopefully after Covid, we will be able to get audited properly." Reef Builders reported that part of an earlier compromise put in place a scheduled ban on wild harvest beginning in 2021. However, some exporters were asking for more time because of the moratorium in combination with COVID shutdowns. "There is no telling how the situation will play out," they wrote. Jake Adams, "There's Another Situation Looming Regarding Wild Corals from Indonesia," Reef Builders, February 18, 2021, https://reefbuilders.com/2021/02/18/theres-another-situation-looming-regarding-wild-corals-from-indonesia/.

172 **more like 28andMe:** A. J. Heyward, "Chromosomes of the Coral *Goniopora lobata* (Anthozoa: Scleractinia)," *Heredity* 55 (1985): 269–71, https://doi.org/10.1038/hdy.1985.101.

Chapter 26. The Scale of Tourism

181 **Tourism accounts for 14 percent:** Rita Kennedy, "The Effects of Tourism in the Caribbean," *USA Today*, March 22, 2018, https://traveltips.usatoday.com/effects-tourism-caribbean-63368 .html. "Sandals Heads Coral Reef Recovery Project," DBS Television, March 14, 2019, https:// www.youtube.com/watch?v=P3fCtLyNKKY. Clear Caribbean, http://www.clearcaribbean .org/our-work/current/. Buddy Dive Resort, https://buddydive.com/diving/reef-renewal/. Cozumel Reef Restoration, https://www.cozumelreefrestoration.com/. Cozumel Coral Reef Restoration Program, https://www.ccrrp.org/. Jamaica's Seascape Caribbean, https://www .seascapecarib.com/. Fragments of Hope, Belize, http://fragmentsofhope.org/.

182 **pillars of marine stewardship:** "Moving towards a Circular Economy," Wave of Change, 2021, https://waveofchange.com/action-line/circular-economy/. "Promoting Responsible Seafood," Wave of Change, 2021, https://waveofchange.com/action-line/responsible-seafood/. "Improving Coastal Health," Wave of Change, 2021, https://waveofchange.com/action-line/coastal-health/. DeMarco Williams, "What Makes Iberostar's Sustainability Charge So Special," *Forbes*, October 29, 2018, https://www.forbes.com/sites/forbestravelguide/2018/10/29/what-makes-iberostars -sustainability-charge-so-special/#352ceb08baa2.

183 **Iberostar's first Caribbean property:** "Our History," Iberostar Group, 2018, https://www .grupoiberostar.com/en/family/history.

183 **Dominican Coastal Restoration Consortium:** Camilo Cortés, "Coral Nurseries Evaluation Season in Dominican Republic Continues," *Dominican Today*, January 15, 2020, https://domini cantoday.com/dr/economy/2020/01/15/coral-nurseries-evaluation-season-in-dominican -republic-continues/.

184 **Coral Vita:** http://www.coralvita.co/. Gator Halpern and Sam Teicher, "Innovation Prize Awardees," J. M. Kaplan Fund, 2017, https://www.jmkfund.org/awardee/gator-halpern-sam -teicher/. Kelvey Vander Hart, "30 Under 30 Highlight: Sam Teicher of Coral Vita," May 14, 2019, American Conservation Coalition, https://www.acc.eco/blog/samteicher30under30.

185 **land-restoration space:** "Restoring Degraded Land to Benefit People and Planet," World Resources Institute, July 14, 2016, https://www.youtube.com/watch?v=J7077absZfg. "Four Things You Need to Know about Forests and Health," United Nations Department of Economic and Social Affairs, https://www.un.org/en/desa/four-things-you-need-know-about -forests-and-health.

185 **efforts to heal lands:** Todd BenDor et al., "Estimating the Size and Impact of the Ecological Restoration Economy," *PloS ONE* 10, no. 6 (2015): e0128339, https://journals.plos.org /plosone/article?id=10.1371/journal.pone.0128339.

Chapter 27. Corals on Ice

187 **Iberostar's lab:** "Iberostar's New Coral Lab Opens for World Oceans Day, Offering Hope for the Future of Caribbean Reefs and Oceans Everywhere," Business Wire, June 6, 2019, https:// www.businesswire.com/news/home/20190606005899/en/Iberostar%E2%80%99s-New -Coral-Lab-Opens-World-Oceans.

188 **interest in cryobiology:** Kenneth B. Storey and Janet M. Storey, "Frozen and Alive," *Scientific American*, December 1990, https://www.scientificamerican.com/article/frozen-and-alive/.

189 **frozen repository:** Mary Hagedorn et al., "Cryopreservation as a Tool for Reef Restoration: 2019," in P. Comizzoli et al., eds., *Reproductive Sciences in Animal Conservation*, 2nd ed. (Cham, Switzerland: Springer, 2019), https://doi.org/10.1007/978-3-030-23633-5_16.

190 **test-tube-born larvae:** Mary Hagedorn et al., "Successful Demonstration of Assisted Gene Flow in the Threatened Coral *Acropora palmata* across Genetically-Isolated Caribbean Populations Using Cryopreserved Sperm," *bioRxiv*, December 10, 2018, https://www.biorxiv.org/content /10.1101/492447v1.full.pdf. Mary Hagedorn et al., "Producing Coral Offspring with Cryopreserved Sperm: A Tool for Coral Reef Restoration," *Scientific Reports* 7, no. 14432 (2017), https:// www.nature.com/articles/s41598-017-14644-x.

190 **a little note:** "Spawning Hope," Smithsonian National Zoo, YouTube, June 15, 2020, https:// www.youtube.com/watch?v=3Bko2bhQgG0.

192 **the gold-and-laser process:** Hagedorn et al., "Cryopreservation as a Tool for Reef Restoration: 2019," in P. Comizzoli et al., eds., *Reproductive Sciences in Animal Conservation, Advances in Experimental Medicine and Biology* 1200 (Cham, Switzerland, Springer, 2019): 489–505, https://doi.org/10.1007/978-3-030-23633-5_16.

Chapter 28. A Coral Named Romeo

193 **FUNDEMAR:** Fundación Dominicana de Estudios Marinos, Inc., https://www.fundemardr.org/.

195 **hermaphrodites:** J. E. N. Veron, M. G. Stafford-Smith, E. Turak, and L. M. DeVantier, *Corals of the World: Overview of Coral Taxonomy*, 2016, Version 0.01 (Beta), accessed July 6, 2021, http://www.coralsoftheworld.org/page/overview-of-coral-taxonomy/?version=0.01.

195 **XXs or XYs:** Takahiro Taguchi et al., "Molecular Cytogenetic Analysis of the Scleractinian Coral *Acropora solitaryensis* Veron & Wallace 1984," *Zoological Science* 31, no. 2 (2014): 89–94, https://bioone.org/journals/zoological-science/volume-31/issue-2/zsj.31.89/Molecular-Cytogenetic-Analysis-of-the-Scleractinian-Coral-Acropora-solitaryensis-Veron/10.2108/zsj.31.89.short.

195 **self-fertilization:** There are marine animals, especially parasites and those that live in deep water, that are able to self-fertilize. Joseph Heller, "Hermaphroditism in Molluscs," *Biological Journal of the Linnean Society* 48, no. 1 (1993): 19–42, https://www.sciencedirect.com/science/article/abs/pii/S0024406683710035. Kirk S. Zigler et al., "Sea Urchin Bindin Divergence Predicts Gamete Compatibility," *Evolution* 59, no. 11 (2005): 2399–404, https://www.bio.fsu.edu/~levitan/publication_pdfs/Zigler%20et%20al%2005.pdf.

196 **sequential hermaphrodites:** "Gender-Bending Fish," Understanding Evolution, University of California Museum of Paleontology, 2021, https://evolution.berkeley.edu/evolibrary/article/fishtree_07. Neil J. Gemmell et al., "Chapter Three—Natural Sex Change in Fish," *Current Topics in Developmental Biology* 134 (2019): 71–117, https://www.sciencedirect.com/science/article/pii/S0070215318301145. Anders Berglund, "Sex Change by a Polychaete: Effects of Social and Reproductive Costs," *Ecology* 67, no. 4 (1986): 837–45, https://www.sciencedirect.com/science/article/abs/pii/S0003347205804051.

Chapter 29. Synchrony of Spawn

202 **the Taíno, who trace:** Earlier estimates proposed that three million people lived on the island when the Spanish arrived, but new research suggests the population was in the tens of thousands. Regardless, only hundreds survived. Andrew Lawler, "Invaders Nearly Wiped Out Caribbean's First People Long Before Spanish Came, DNA Reveals," *National Geographic*, December 23, 2020, https://www.nationalgeographic.com/history/2020/12/invaders-nearly-wiped-out-caribbeans-first-people-long-before-spanish-came-dna-reveals/. Robert M. Poole, "What Became of the Taíno?," *Smithsonian*, October 2011, https://www.smithsonianmag.com/travel/what-became-of-the-taino-73824867/. Abdul Rob, "Taíno: Indigenous Caribbeans," Black History Month, December 2, 2016, https://www.blackhistorymonth.org.uk/article/section/pre-colonial-history/taino-indigenous-caribbeans/.

202 **Columbus wrote:** Kirkpatrick Sale, *The Conquest of Paradise: Christopher Columbus and the Columbian Legacy* (New York: Knopf, 1990), 100, https://archive.org/details/conquestofpa00sale/page/100/mode/2up.

203 **Grupo Puntacana:** "Our History and the PUNTACANA Brand," Puntacana Resort & Club, https://www.puntacana.com/assets/our-history-and-the-puntacana-brand.pdf.

204 **the country's economy:** Ana M. López, "Dominican Republic: Travel and Tourism Participation in the GDP by Type 2010–2019," Statista, December 9, 2020, https://www.statista.com/statistics/874522/dominican-republic-travel-tourism-total-contribution-to-gdp-by-share/. Ana M. López, "Expenditures of International Tourists in the Dominican Republic from 2010 to 2019," Statista, 2020, https://www.statista.com/statistics/814782/dominican-republic-tourism-revenue/.

204 **Jake Kheel:** Jake Kheel, *Waking the Sleeping Giant* (Lioncrest, 2021), https://jakekheel.com/.

204 **Fundación Grupo Puntacana:** Fundación Grupo Puntacana, https://puntacana.org/en/.

206 **Horniman Museum and Gardens:** "Project Coral," Horniman Museum & Gardens, https://www.horniman.ac.uk/about/project-coral. Philip Henry Gosse, who wrote *The Aquarium: An Unveiling of the Wonders of the Deep Sea* in 1854 (in which he coined the word *aquarium*) and *A Handbook to the Marine Aquarium* in 1855, built the Horniman's aquaria in 1903.

208 **coral spawning systems:** Jamie Craggs et al., "Inducing Broadcast Coral Spawning Ex Situ: Closed System Mesocosm Design and Husbandry Protocol," *Ecology and Evolution* 7, no. 24 (2017): 11066–78, https://onlinelibrary.wiley.com/doi/full/10.1002/ece3.3538. "Jamie Craggs: Overview of 5 Years Spawning Coral in Captivity. | MACNA 2018," BRStv (Bulk Reef Supply), YouTube, March 25, 2019, https://www.youtube.com/watch?v=q81mA8rdH7A.

208 **spawned a new species:** Morgan McFall-Johnsen, "Scientists Reproduced Atlantic Coral in a Laboratory for the First Time. Aquarium-Grown Coral Could Save America's 'Great Barrier Reef,'" *Business Insider*, August 28, 2019, https://www.businessinsider.com/atlantic-coral-reproduced-florida-laboratory-first-time-2019-8?op=1. Lauren M. Johnson, "A Scientific Breakthrough at the Florida Aquarium Could Save 'America's Great Barrier Reef,'" CNN.com, August 22, 2019, https://www.cnn.com/2019/08/21/us/historic-coral-discovery-scn-trnd/index.html. Jason Bittel, "The Fate of This Coral Species Rests in a Dark Room in Tampa," onEarth, Natural Resources Defense Council, October 28, 2019, https://www.nrdc.org/onearth/fate-coral-species-rests-dark-room-tampa.

209 **Only fifty colonies:** Karen L. Neely and Cynthia L. Lewis, "Rapid Population Decline of the Pillar Coral *Dendrogyra cylindrus* along the Florida Reef Tract," *bioRxiv*, May 9, 2020, https://www.biorxiv.org/content/10.1101/2020.05.09.085886v1.full.pdf.

209 **spawning another species:** Alaa Elassar, "The Florida Aquarium Just Made a Breakthrough That Will Help Save the Third Largest Coral Reef in the World," CNN, April 23, 2020, https://www.cnn.com/2020/04/22/us/florida-aquarium-first-reproduce-ridhed-cactus-coral-trnd/index.html.

209 **coral's life cycle:** Jamie Craggs et al., "Completing the Life Cycle of a Broadcast Spawning Coral in a Closed Mesocosm," *Invertebrate Reproduction & Development* 64, no. 3 (2020): 244–47, https://www.tandfonline.com/doi/abs/10.1080/07924259.2020.1759704.

Chapter 30. Coral Kindergarten

210 **coral babies didn't make it:** Alasdair J. Edwards et al., "Direct Seeding of Mass-Cultured Coral Larvae Is Not an Effective Option for Reef Rehabilitation," *Marine Ecology Progress Series* 525 (2015): 105–16, http://www.int-res.com/abstracts/meps/v525/p105-116/.

211 **SECORE International:** SECORE International, http://www.secore.org/site/home.html.

213 **"No risk is more terrifying":** Hope Jahren, *Lab Girl* (New York: Alfred A. Knopf, 2016), 52.

213 **what convinces a coral larva:** Marie E. Strader et al., "Molecular Characterization of Larval Development from Fertilization to Metamorphosis in a Reef-Building Coral," *BMC Genomics* 19, no.17 (2018), https://doi.org/10.1186/s12864-017-4392-0. Mark J. A. Vermeij et al., "Coral Larvae Move toward Reef Sounds," *PLoS ONE* 5, no. 5 (2010): e10660, https://doi.org/10.1371/journal.pone.0010660. Taylor N. Whitman et al., "Settlement of Larvae from Four Families of Corals in Response to a Crustose Coralline Alga and Its Biochemical Morphogens," *Scientific Reports* 10 (2020): 16397, https://doi.org/10.1038/s41598-020-73103-2.

214 **coral kindergarten experiment:** Valérie F. Chamberland et al., "New Seeding Approach Reduces Costs and Time to Outplant Sexually Propagated Corals for Reef Restoration," *Scientific Reports* 7, no. 18076 (2017), https://www.nature.com/articles/s41598-017-17555-z.

Chapter 31. Category 5

216 **likely as a hurricane:** "Hurricane Dorian: Path of Destruction," BBC News, September 9, 2019, https://www.bbc.com/news/world-latin-america-49553770.

217 **heat is the fuel:** Rebecca Hersher, "Expect More Tropical Storms, NOAA Warns," NPR.org, April 9, 2021, https://www.npr.org/2021/04/09/985804008/expect-more-tropical-storms-noaa-warns. Thomas R. Knutson et al., "ScienceBrief Review: Climate Change Is Probably Increasing the Intensity of Tropical Cyclones," *Critical Issues in Climate Change Science*, March 26, 2021, https://news.sciencebrief.org/cyclones-mar2021. Brian K. Sullivan, "Warmest Oceans on Record Could Set Off a Year of Extreme Weather," American Institute of Physics, April 18, 2020, https://phys.org/news/2020-04-warmest-oceans-year-extreme-weather.html. "Assessing the Global Climate in 2020," NOAA Centers for Environmental Information, January 14, 2021, https://www.ncei.noaa.gov/news/global-climate-202012

217 **official death toll:** Jacqueline Charles, "Bahamas Has No Idea of Dorian Death Toll after Names of the Missing Were Removed from List," *The Miami Herald*, June 12, 2020, https://www.miamiherald.com/news/nation-world/world/americas/article243472646.html

#storylink=cpy. Nicole Chavez and Chandler Thornton, "600 People Are Still Missing in the Bahamas Weeks after Hurricane Dorian," CNN, September 27, 2019, https://www.cnn.com /2019/09/27/americas/bahamas-hurricane-dorian-missing/index.html.

218 **Infrastructure damage:** Katherine Chiglinsky, "Hurricane Dorian Seen Costing Bahamas at Least $7 Billion," Bloomberg, September 5, 2019, https://www.bloomberg.com/news/articles /2019-09-05/hurricane-dorian-seen-costing-the-bahamas-at-least-7-billion.

218 **attention back to corals:** Brita Belli, "After Dorian, Saving Coral Reefs Even More Urgent for This Alumni Startup," *Yale News,* October 30, 2019, https://news.yale.edu/2019/10/30/after -dorian-saving-coral-reefs-even-more-urgent-alumni-startup.

219 **"There is hope":** Craig Dahlgren and Krista Sherman, "Preliminary Assessment of Hurricane Dorian's Impact on Coral Reefs of Abaco and Grand Bahama," *International Coral Reef Initiative and Perry Institute of Marine Science,* February 8, 2020, p. 3, https://www.icriforum.org/prelimi nary-assessment-of-hurricane-dorians-impact-on-coral-reefs-of-abaco-and-grand-bahama/. Chris D'Angelo, "Fearing Devastation, Coral Scientists Begin Post-Dorian Reef Surveys," *Huff-Post,* October 8, 2019, https://www.huffpost.com/entry/coral-reefs-bahamas-hurricane-dorian _n_5d9cd845e4b02c9da03fef03?ncid=engmodushpmg00000004.

Chapter 32. X-tinguished

223 **Matt Mulrennan:** Matt has since taken a position with the environmental investment event EnVest, https://envest.earth/.

223 **John Steinbeck:** Rose Pastore, "John Steinbeck's 1966 Plea to Create a NASA for the Oceans," *Popular Science,* May 20, 2014, https://www.popsci.com/article/technology/john-steinbecks-1966 -plea-create-nasa-oceans/.

224 **funding for the oceans:** "Line Office Summary," *NOAA Budget Summary,* 2021, 55–56, https:// www.noaa.gov/sites/default/files/legacy/document/2020/Mar/508%20Compliant_NOAA%20 FY21%20Budget%20Blue%20Book%20Summary.pdf. "Final FY21 Appropriations: NASA," American Institute of Physics, February 11, 2021, https://www.aip.org/fyi/2021/final-fy21 -appropriations-nasa. Michael Conathan, "Rockets Top Submarines: Space Exploration Dollars Dwarf Ocean Spending," Center for American Progress, June 18, 2013, https://www.american progress.org/issues/green/news/2013/06/18/66956/rockets-top-submarines-space-exploration -dollars-dwarf-ocean-spending/.

224 **"Life Below Water":** "Sustainable Development Goals, Showing Results for 'Life Below Water,'" SDGFunders, Candid, 2021, https://sdgfunders.org/sdgs/dataset/recent/goal/life -below-water/.

224 **report released:** "Based on findings from simulation modelling, SR15 concluded that 'coral reefs are projected to decline by a further 70–90% at 1.5°C (very high confidence) with larger losses (>99%) at 2°C (very high confidence).'" N. L. Bindoff et al., "Changing Ocean, Marine Ecosystems, and Dependent Communities," in H.-O. Pörtner et al., eds., *IPCC Special Report on the Ocean and Cryosphere in a Changing Climate,* 2019, in press, https://www.ipcc.ch/srocc/chapter /chapter-5/.

226 **Glowing Glowing Gone:** Larissa Faw, "World Surf League 'Glows' for Coral Reef Initia-tive," *MediaPost,* August 20, 2019, https://www.mediapost.com/publications/article/339523 /world-surf-league-glows-for-coral-reef-initiativ.html. "Getting Creative with Coral Reef Conservation," United Nations Environment Programme, November 5, 2019, https://www .unenvironment.org/news-and-stories/story/getting-creative-coral-reef-conservation. "UNEP Launches Glowing Glowing Gone Campaign on Loss of Coral Due to Climate Change," United Nations Environment Programme, February 28, 2020, https://www.unenvironment.org/news -and-stories/story/unep-launches-glowing-glowing-gone-campaign-loss-coral-due-climate -change.

227 **Mission: Iconic Reefs:** "Restoring Seven Iconic Reefs: A Mission to Recover the Coral Reefs of the Florida Keys," NOAA Fisheries, 2021, https://www.fisheries.noaa.gov/southeast/habitat -conservation/restoring-seven-iconic-reefs-mission-recover-coral-reefs-florida-keys.

228 **Times Square:** "Hope for Coral Reefs Takes Hold in Times Square," Coral Restoration Founda-tion, October 25, 2019, https://www.coralrestoration.org/post/hope-for-coral-reefs-takes-hold -in-times-square.

Chapter 33. The Lifeboats

229 **50 Reefs:** "50 Reefs: A Global Plan to Save Coral Reefs," The Ocean Agency, 2020, https://www.50reefs.org/. Hawthorne L. Beyer et al., "Risk-Sensitive Planning for Conserving Coral Reefs under Rapid Climate Change," *Conservation Letters*, June 27, 2018, https://conbio.online library.wiley.com/doi/full/10.1111/conl.12587. Ove Hoegh-Guldberg et al., "Securing a Long-term Future for Coral Reefs," *Trends in Ecology & Evolution* 33, no. 12 (2018): 936–44, https://www.bbhub.io/dotorg/sites/2/2018/10/Securing-a-Long-term-Future-for-Coral-Reefs.pdf.

229 **economist Harry Markowitz:** "Prize in Economic Sciences 1990," The Nobel Prize, The Royal Swedish Academy of Sciences, October 16, 1990, https://www.nobelprize.org/prizes/economic -sciences/1990/press-release/.

230 **fifty-first reef:** Jessica Leber, "Inside the Mission to Find 50 Reefs That Could Survive Climate Change," *The New Humanitarian*, July 26, 2018, https://www.newsdeeply.com/oceans/articles /2018/07/26/inside-the-mission-to-find-50-reefs-that-could-survive-climate-change.

230 **"lifeboat ethics":** Garrett Hardin, "Living on a Lifeboat," *BioScience* 24, no. 10, October 1974, https://www.garretthardinsociety.org/articles_pdf/living_on_a_lifeboat.pdf.

231 **Allen Coral Atlas:** Allen Coral Atlas, https://www.allencoralatlas.org/. Eillie Anzilotti, "Inside the Effort to Map the World's Dying Coral Reefs from Space," *Fast Company*, July 19, 2019, https://www.fastcompany.com/90378706/inside-the-effort-to-map-the-worlds-dying-coral -reefs-from-space. Mark Phillips, "Scientists Use Mini-Satellites in Effort to Save the World's Coral Reefs," CBS News, March 11, 2019, https://www.cbsnews.com/news/great-barrier-reef -scientists-use-mini-satellites-effort-to-save-coral-reefs-2019-03-11/. Adele Peters, "This Map Is the First-ever to Show Every Single Coral Reef in the Caribbean," *Fast Company*, December 9, 2020, https://www.fastcompany.com/90580296/this-map-is-the-first-ever-to-show-every-single -coral-reef-in-the-caribbean. Other groups also are working on mapping coral reefs with remote sensing, including a significant project at Khaled bin Sultan Living Oceans Foundation: Sam J. Purkis et al., "High-Resolution Habitat and Bathymetry Maps for 65,000 Sq. Km of Earth's Remotest Coral Reefs," *Coral Reefs* 38 (2019): 467–88, https://link.springer.com/content/pdf /10.1007%2Fs00338-019-01802-y.pdf.

232 **Helen's graduate work:** Helen E. Fox et al., "Rebuilding Coral Reefs: Success (and Failure) 16 Years after Low-Cost, Low-Tech Restoration," *Restoration Ecology* 27, no. 4 (2019): 862–69, https://onlinelibrary.wiley.com/doi/abs/10.1111/rec.12935.

232 **complicated collaboration:** The field operations role has since transitioned to Arizona State University; Helen now serves as expert adviser to the Coral Atlas from her position as conservation science director of the Coral Reef Alliance, https://coral.org/our-team/.

Chapter 34. Reef Investments

237 **Rob Weary:** Rob has since started his own company, Aqua Blue Investments LLC, working on debt conversions to finance marine conservation.

237 **Tropical Forest Conservation Act:** "Tropical Forest Conservation Act," The Nature Conservancy, April 6, 2020, https://www.nature.org/en-us/about-us/who-we-are/how-we-work/policy /tropical-forest-conservation-act/.

239 **"a great commitment":** "Event Spotlight: Good News Out of Rio+20: Six Heads of State Commit to Invest in Nature," The Nature Conservancy and Global Island Partnership, June 21, 2012, http://www.glispa.org/images/glispa/events/GLISPA_EventSpotlight_Rio20_print.pdf. "Seychelles Swaps Debt for Nature," The Economist Group World Ocean Initiative, April 8, 2020, https://ocean.economist.com/blue-finance/articles/seychelles-swaps-debt-for-nature.

239 **bond for marine conservation:** On blue bonds and green bonds, see Lyubov Pronina, "What Are Green Bonds and How 'Green' Is Green?" Bloomberg, March 24, 2019, https://www .bloomberg.com/news/articles/2019-03-24/what-are-green-bonds-and-how-green-is-green -quicktake. Troy Segal, "Green Bond," Investopedia, updated April 24, 2021, https://www .investopedia.com/terms/g/green-bond.asp. "Seychelles Launches World's First Sovereign Blue Bond," The World Bank, October 29, 2018, https://www.worldbank.org/en/news/press-release /2018/10/29/seychelles-launches-worlds-first-sovereign-blue-bond. Navindu Katugampola and Matthew Slovik, "The New Sustainability Bond That Aims to Protect Oceans," Morgan Stanley

Institute for Sustainable Investing, September 18, 2019, https://www.morganstanley.com/ideas /blue-bonds-sustainable-investing-next-wave.

240 **blue bonds to unlock investment:** "Sovereign Blue Bond Issuance: Frequently Asked Questions," The World Bank, October 29, 2018, https://www.worldbank.org/en/news/feature/2018 /10/29/sovereign-blue-bond-issuance-frequently-asked-questions.

240 **thirteen marine protected areas:** Malavika Vyawahare, "Seychelles Extends Protection to Marine Area Twice the Size of Great Britain," *Mongabay*, March 30, 2020, https://news.mongabay .com/2020/03/seychelles-extends-protection-to-marine-area-twice-the-size-of-great-britain/. "Seychelles Hits 30% Marine Protection Target after Pioneering Debt Restructuring Deal," The Nature Conservancy, March 26, 2020, https://www.nature.org/en-us/newsroom/seychelles -achieves-marine-protection-goal/.

241 **Coastal wetlands:** Carlos M. Duarte et al., "Rebuilding Marine Life," *Nature* 580 (2020): 39–51, https://doi.org/10.1038/s41586-020-2146-7.

241 **"insurance to protect":** YCC Team, "Cancun Businesses Take Out Insurance Policy on a Coral Reef," Yale Climate Connections, January 3, 2020, https://www.yaleclimateconnections.org /2020/01/cancun-businesses-take-out-insurance-policy-on-a-coral-reef/.

241 **the insurance policy was issued:** Sophie Hares, "Mexican Coral Reef and Beach Get Unique Insurance Policy against Hurricane Damage," Reuters, March 8, 2018, https://www.reuters.com /article/us-mexico-environment-reefs/mexican-coral-reef-and-beach-get-unique-insurance -policy-against-hurricane-damage-idUSKCN1GK384. "Designing a New Type of Insurance to Protect the Coral Reefs, Economies and the Planet," Swiss Re, December 10, 2019, https://www .swissre.com/our-business/public-sector-solutions/thought-leadership/new-type-of-insurance -to-protect-coral-reefs-economies.html. Mark Tercek, "Business to the Rescue! Insurance for Reef Restoration," The Nature Conservancy, March 8, 2018, https://www.nature.org/en-us /about-us/who-we-are/our-people/mark-tercek/business-to-the-rescue—insurance-for-reef -restoration/.

242 **The 2020 hurricane season:** Catrin Einhorn and Christopher Flavelle, "A Race against Time to Rescue a Reef from Climate Change," *The New York Times*, December 5, 2020, https://www .nytimes.com/2020/12/05/climate/Mexico-reef-climate-change.html.

243 **"public-private partnership":** Benjamin K. Sovacool, "An International Comparison of Four Polycentric Approaches to Climate and Energy Governance," *Energy Policy* 39, no. 6 (2011): 3832–44, https://doi.org/10.1016/j.enpol.2011.04.014. Veronica Scotti, "Public-Private Partnerships Are the Key to Tackling Climate Change. Here's Why," World Economic Forum, January 14, 2020, https://www.weforum.org/agenda/2020/01/in-the-fight-against-climate-change-public -private-partnerships-are-the-only-way-to-go/.

Chapter 35. Cancellation

247 *A Farewell to Ice*: Peter Wadhams, *A Farewell to Ice: A Report from the Arctic* (New York: Oxford University Press, 2017).

249 **geoengineer our way out:** Andrew Revkin, "Scientists Focused on Geoengineering Challenge the Inevitability of Multi-millennial Global Warming," *Medium*, September 2, 2016, https:// revkin.medium.com/geoengineering-proponents-challenge-the-inevitability-of-multi -millennial-global-warming-cef6e54b365c.

253 **high bleaching probability:** Renee Cluff, "Reef Scientists on Alert for Great Barrier Reef Bleaching," *Tropic Now*, February 18, 2020, https://www.tropicnow.com.au/2020/february/18 /reef-scientists-on-alert-for-great-barrier-reef-bleaching.html.

253 **"Alarm bells":** Graham Readfearn, "Great Barrier Reef Could Face 'Most Extensive Coral Bleaching Ever,' Scientists Say," *The Guardian*, February 21, 2020, https://www.theguardian .com/environment/2020/feb/22/great-barrier-reef-could-face-most-extensive-coral-bleaching -ever-scientists-say. Maddie Stone, "The Great Barrier Reef Is Heading for a Mass Bleaching of Unprecedented Scale," *Vice*, March 4, 2020, https://www.vice.com/en_us/article/y3mxmg/great -barrier-reef-coral-bleaching-2020.

Chapter 36. Collapse

259 **Terry Hughes:** @ProfTerryHughes, Twitter, March 26, 2020, 5:06 a.m., https://twitter.com /ProfTerryHughes/status/1243117101788721153.

259 **2020 Great Barrier Reef bleaching:** "Coral Bleaching Events," Australian Institute of Marine Science, 2017, https://www.aims.gov.au/docs/research/climate-change/coral-bleaching/bleaching-events.html.

261 **exclusive story:** Graham Readfearn, "Scientists Trial Cloud Brightening Equipment to Shade and Cool Great Barrier Reef," *The Guardian*, April 16, 2020, https://www.theguardian.com/environment/2020/apr/17/scientists-trial-cloud-brightening-equipment-to-shade-and-cool-great-barrier-reef.

262 **Traditional Owners:** "Traditional Owners of the Great Barrier Reef," Great Barrier Reef Marine Park Authority, 2021, https://www.gbrmpa.gov.au/our-partners/traditional-owners/traditional-owners-of-the-great-barrier-reef.

262 **carbon emissions fell:** E&T editorial staff, "Largest CO_2 Emissions Drop in History Due to Covid-19 Pandemic," *Engineering & Technology*, October 15, 2020, https://eandt.theiet.org/content/articles/2020/10/covid-19-pandemic-caused-largest-co2-drop-in-history/.

Chapter 37. Flicker

263 **Breonna Taylor:** Darcy Costello and Tessa Duvall, "Who Are the Louisville Officers Involved in the Breonna Taylor Shooting? What We Know," *Louisville Courier Journal*, October 16, 2020, https://www.courier-journal.com/story/news/politics/metro-government/2020/05/16/breonna-taylor-shooting-what-we-know-louisville-police-officers-involved/5200879002/.

264 **systemic racism:** Niall McCarthy, "Police Shootings: Black Americans Disproportionately Affected," *Forbes*, May 28, 2020, https://www.forbes.com/sites/niallmccarthy/2020/05/28/police-shootings-black-americans-disproportionately-affected-infographic/#3e1a0a5059f7. John Gramlich, "Black Imprisonment Rate in the U.S. Has Fallen by a Third since 2006," Pew Research Center, May 6, 2020, https://www.pewresearch.org/fact-tank/2020/05/06/share-of-black-white-hispanic-americans-in-prison-2018-vs-2006/.

264 **string of hashtags:** Carly Mallenbaum, "#BlackBirdersWeek, #BlackInNeuro: Black Scientists, Physicians Are Using Hashtags to Uplift," *USA Today*, August 4, 2020, https://www.usatoday.com/story/life/2020/08/04/blackinneuro-blackinchem-can-hashtags-help-black-scientists-build-community-spotlight-excellence/5541431002/. "#ShutdownAcademia #ShutdownSTEM," June 10, 2020, https://www.shutdownstem.com/.

264 **Black professors in evolutionary science:** Joseph L. Graves Jr., "African Americans in Evolutionary Science: Where We Have Been, and What's Next," *Evolution: Education and Outreach* 12, no. 18 (2019), https://link.springer.com/article/10.1186/s12052-019-0110-5. Bas Hofstra et al., "The Diversity–Innovation Paradox in Science," *Proceedings of the National Academies of Sciences of the United States* 117, no. 17 (2020): 9284–91, https://www.pnas.org/content/pnas/117/17/9284.full.pdf.

266 **essay by Cinda P. Scott:** Cinda P. Scott, "Changing the Narrative in Conservation and Marine Sciences Is the Responsibility of Us All," *The Waterlust Blog*, July 21, 2020, https://waterlust.com/blogs/the-waterlust-blog/changing-the-narrative.

266 **accumulation of barriers:** In 2016, the total number of science and engineering graduate students in the United States was 620,489. There were 2,743 in ocean sciences, and 18 were Black or African American women. S&E graduate students, by field, sex, citizenship, ethnicity, and race: 2016 Table 3-1, "Women, Minorities, and Persons with Disabilities in Science and Engineering," National Center for Science and Engineering Statistics (NCSES), National Science Foundation, March 8, 2019, https://ncses.nsf.gov/pubs/nsf19304/data.

Chapter 38. Survival Genes

270 **Eric Fisher:** "Meet the GBR Biology Team," GBR Biology, https://www.gbrbiology.com/our-team/.

271 **random mutations:** Kashmira Gander, "First 'Significant' Coronavirus Mutation Discovered in Preliminary Study," *Newsweek*, April 14, 2020, https://www.newsweek.com/coronavirus-mutation-study-covid-19-1497745.

271 **virtual tour of SeaSim:** "National Sea Simulator," Australian Institute of Marine Science, https://www.aims.gov.au/seasim.

273 **genetic variability:** Mikhail V. Matz et al., "Potential and Limits for Rapid Genetic Adaptation to Warming in a Great Barrier Reef Coral," *PLoS Genetics* 14, no. 4 (2018): e1007220, https://doi.org/10.1371/journal.pgen.1007220.

273 **in the prestigious journal *Science*:** Zachary L. Fuller et al., "Population Genetics of the Coral *Acropora millepora*: Toward Genomic Prediction of Bleaching," *Science* 369, no. 6501 (2020), https://science.sciencemag.org/content/369/6501/eaba4674.full.

277 **disproportionately affected:** William Wan, "Coronavirus Kills Far More Hispanic and Black Children Than White Youths, CDC Study Finds," *The Washington Post*, September 15, 2020, https://www.washingtonpost.com/health/2020/09/15/covid-deaths-hispanic-black-children/. Justin Worland, "Why the Larger Climate Movement Is Finally Embracing the Fight against Environmental Racism," *Time*, July 9, 2020, https://time.com/5864704/environmental-racism-climate-change/.

Chapter 39. Brightening

281 **a giant dust storm:** Jeff Masters, "Saharan Dust Cloud Was Most Intense in Decades, and More, Though Milder, Are Coming," Yale Climate Connections, June 29, 2020, https://www.yaleclimate connections.org/2020/06/saharan-dust-cloud-was-most-intense-in-decades/. Rob Gutro, "NASA-NOAA's Suomi NPP Satellite Analyzes Saharan Dust Aerosol Blanket," NASA, June 26, 2020, https://www.nasa.gov/feature/goddard/2020/nasa-noaa-s-suomi-npp-satellite-analyzes -saharan-dust-aerosol-blanket.

Chapter 40. The Coin Toss

284 **Global Fund for Coral Reefs:** Global Fund for Coral Reefs, https://globalfundcoralreefs.org/. "Global Fund for Coral Reefs, High-Level Event—September 16th, 2020," Global Fund for Coral Reefs, September 22, 2020, https://www.youtube.com/watch?v=xbtmo4004B0. Francis Staub, "How the World Is Coming Together to Save Coral Reefs," World Economic Forum, December 4, 2020, https://www.weforum.org/agenda/2020/12/how-the-world-is-coming -together-to-save-coral-reefs/.

284 **BNP Paribas:** Based in France, it is the world's eighth-largest bank by total assets. With a tagline, "The bank for a changing world," it touts a strong strategy in line with the United Nation's Sustainable Development Goals. However, since the signing of the Paris Climate Accords, BNP has financed roughly $120 billion in fossil fuel projects, more than any other bank in the EU. Damian Carrington, "Big Banks' Trillion-Dollar Finance for Fossil Fuels 'Shocking,' Says Report," *The Guardian*, March 24, 2021, https://www.theguardian.com/environment/2021/mar /24/big-banks-trillion-dollar-finance-for-fossil-fuels-shocking-says-report.

284 **Prince Albert II of Monaco Foundation:** Fondation Prince Albert II de Monaco, https://www.fpa2.org/home.html.

287 **UK minister for Pacific and the Environment:** "The Rt Hon Lord Zac Goldsmith," Gov.UK, https://www.gov.uk/government/people/zac-goldsmith.

288 **pandemic as a catalyst:** Somini Sengupta, "Economic Giants Are Restarting. Here's What It Means for Climate Change," *The New York Times*, May 29, 2020, https://www.nytimes.com /2020/05/29/climate/coronavirus-economic-stimulus-climate.html. Jordan Davidson, "BP to Cut Oil and Gas Production 40%, Invest 10x More in Green Energy," EcoWatch, August 5, 2020, https://www.ecowatch.com/bp-green-energy-investment-2646892538.html. Rebecca Shabad and Geoff Bennett, "Biden Signs Executive Actions on Climate Change: 'It's Time to Act,'" NBC News, January 27, 2021, https://www.nbcnews.com/politics/white-house/biden -sign-executive-actions-climate-change-n1255814.

288 **In our seas:** Rachel Fritts, "US Plans to Protect Thousands of Miles of Coral Reefs in Pacific and Caribbean," *The Guardian*, December 4, 2020, https://www.theguardian.com/environ ment/2020/dec/04/us-pitches-coral-protections-thousands-miles-climate-change-threat. Fiona Harvey, "Global Sustainable Fishing Initiative Agreed by 14 Countries," *The Guardian*, December 2, 2020, https://www.theguardian.com/environment/2020/dec/02/global-sustain able-fishing-initiative-agreed-by-14-countries.

289 **without hope, there is no grief:** I recall hearing this line from Krista Tippett, host of the *On Being* podcast, https://onbeing.org.

291 **I signed up:** The reporting for this trip was originally published in Juli Berwald, "The Gulf's Secret Garden," *Texas Monthly*, September 2021, https://www.texasmonthly.com/travel/behind -the-fight-to-save-the-gulfs-spectacular-coral-reefs/.

291 **tripling its size:** "Sanctuary Expansion," Flower Garden Banks National Marine Sanctuary, National Ocean Service, NOAA, https://flowergarden.noaa.gov/management/sanctuary expansion.html.

291 **Republican from Houston:** Clint Moore, http://clintmoore.com/.

292 **monitoring surveys:** "Coral Cap Species of Flower Garden Banks National Marine Sanctuary," Flower Garden Banks National Marine Sanctuary, https://nmsflowergarden.blob.core .windows.net/flowergarden-prod/media/archive/document_library/aboutdocs/fgbnmscoral capspecies.pdf.

292 *increase* **in coral cover:** Michelle A. Johnston et al., "Persistence of Coral Assemblages at East and West Flower Garden Banks, Gulf of Mexico," *Coral Reefs* 35 (2016): 821–26, https://link .springer.com/content/pdf/10.1007/s00338-016-1452-x.pdf

292 **shore-borne impacts:** Elizabeth Weinberg, "Scientists Unable to Identify 'Smoking Gun' in 2016 Coral Mortality Event at Flower Garden Banks National Marine Sanctuary," NOAA National Marine Sanctuaries, August 2018, https://sanctuaries.noaa.gov/news/aug18/flower-garden -banks-coral-mortality-symposium.html

Index